The U.S. Machine Tool Industry and the Defense Industrial Base

Committee on the Machine Tool Industry
Manufacturing Studies Board
Commission on Engineering and Technical Systems
National Research Council

NATIONAL ACADEMY PRESS
Washington. D.C. 1983

NOTICE: The project that is the subject of this report was approved by the Governing Board of the National Research Council, whose members are drawn from the councils of the National Academy of Sciences, the National Academy of Engineering, and the Institute of Medicine. The members of the committee responsible for the report were chosen for their special competences and with regard for appropriate balance.

This report has been reviewed by a group other than the authors according to procedures approved by a Report Review Committee consisting of members of the National Academy of Sciences, the National Academy of Engineering, and the Institute of Medicine.

The National Research Council was established by the National Academy of Sciences in 1916 to associate the broad community of science and technology with the Academy's purposes of furthering knowledge and of advising the federal government. The Council operates in accordance with general policies determined by the Academy under the authority of its congressional charter of 1863, which establishes the Academy as a private, nonprofit, self-governing membership corporation. The Council has become the principal operating agency of both the National Academy of Sciences and the National Academy of Engineering in the conduct of their services to the government, the public, and the scientific and engineering communities. It is administered jointly by both Academies and the Institute of Medicine. The National Academy of Engineering and the Institute of Medicine were established in 1964 and 1970, respectively, under the charter of the National Academy of Sciences.

This report represents work under contract number DAAK21-82-C-0091 between the U.S. Department of the Army and the National Academy of Sciences.

A limited number of copies are available from:

Manufacturing Studies Board
National Academy of Sciences
2101 Constitution Avenue, N.W.
Washington, D.C. 20418

Printed in the United States of America

COMMITTEE ON THE MACHINE TOOL INDUSTRY, Phase II

JAMES E.ASHTON, Vice President, Rockwell International, <u>Chairman</u>
MARGARET B.W.GRAHAM, School of Management, Boston University, <u>Vice Chairman</u>
ARDEN L.BEMENT, Vice President, Technical Resources, TRW, Incorporated
JOSEPH E.CLANCY, President, Bridgeport Machine
MICHAEL W.DAVIS, President, White-Sundstrand Machine Tool Company
BELA GOLD, Director, Research Program in Industrial Economics, Case Western Reserve University
HAMILTON HERMAN, Management Consultant
NATHANIEL S.HOWE, Senior Vice President, Litton Industries
RICHARD A.JAY, President, Work in Northeast Ohio Council
ROBERT B.KURTZ, Retired Senior Vice President, General Electric
WALTER L.MACORITTO, Manager, Manufacturing Research, Lockheed Corporation
ROGER B.ORLOFF, Vice President, Corporate Finance Group, Girard Bank
HENRY D.SHARPE, Jr., Chairman, Brown and Sharpe Manufacturing Company
CEDRIC L.SUZMAN, Vice President and Educational Program Director, Southern Center for International Studies
JOHN G.T.THORNTON, Editor, Robot Insider
ROBERT TRIMBLE, Vice President, Contracts, Martin Marietta Aerospace
WILLIAM M.TRUSKA, Jr., President, Rousselle Corporation
BERNARD WATERMAN, Industrial College of the Armed Forces
FRED WOHLFAHRT, Project Manager, Technical Management and Systems Consulting, General Electric Company

PREFACE

This report presents the findings and recommendations of the second of two committees formed by the Manufacturing Studies Board at the request of the Department of Defense to study the defense readiness and international competitiveness of the U.S. machine tool industry.

The Committee on the Machine Tool Industry, Phase I, was a three-month effort beginning in October 1981. It reviewed prior studies, defined issues, and designed the study to be undertaken in Phase II, which has resulted in this report.

The Phase I study is a stepping stone for the Phase II analysis—this report —of the machine tool industry's competitiveness and defense readiness, in the light of its changing structure and capabilities. The committee is also indebted to the earlier studies of the machine tool industry, particularly those by the Machine Tool Task Force, the Defense Science Board, and the National Academy of Engineering.

Important trends, however, have intensified in the machine tool industry since the writing of those earlier reports. New process technology is rapidly widening the scope of the industry beyond the traditional concept of metal-cutting and -forming equipment; further, structural changes within the industry include an increasing number of mergers, acquisitions, and joint ventures. A new study was needed that took as its starting point an emerging machine tool capability, reflecting both technological and structural changes.

The Phase I committee designed a study to interpret the significance of this transformation of the industry for the Department of Defense (DOD) and to make recommendations for policy based on DOD's needs and the emerging industry. This report presents the results of that study.

The Committee on the Machine Tool Industry, Phase II, is solely responsible for this report. A number of others, though, have made invaluable contributions. Primary among these is the Phase I committee, whose definition of issues and study design were the basis for the work represented in this report. Mel Horwitch generously gave many hours to assist the Committee in identifying trends and policy options; he also provided valuable comments on the several drafts of the report. Study directors Joel Goldhar (first half) and George Kuper (second half) contributed many of the insights to the Committee's discussion. Consultants Stephen Merrill and Jack Bloom provided analyses of the relationships among DOD, the machine tool industry, and prime contractors, and did the initial drafts of sections of the report. Staff officer Janice Greene assisted in the Committee's analysis and writing. Consultant William Levitt conducted and analyzed a survey of recent machine tool purchases by domestic firms. Consultant Harold Davidson provided a wealth of historical and procedural information from the Department of Defense. Charles Downer, of the National Machine Tool Builders' Association, was another important source of data. Consultant Edgar Weinberg provided statistical backup. Consultant George Krumbhaar researched issues pertaining to the viewpoints of prime contractors and also organized the Committee's comments into the final version of this report. Consultant Deborah Tomusko conducted case studies at machine tool builders; her research forms the basis for parts of Chapter 2. Staff associate Georgene Menk was responsible for the administrative work of the Committee, and Donna Reifsnider and Frances Shaw ably typed this report.

The aforementioned help notwithstanding, there would be no report without the diligent efforts of a very hard-working volunteer committee, including a talented group of drafters led by our Vice Chairperson, Margaret Graham.

James E. Ashton

CONTENTS

1. **INTRODUCTION**   1
   - Statement of the Problem   1
   - The U.S. Machine Tool Industry: The Problems of Maturity   2
   - The DOD Interest   3
   - Approach to the Study   4
   - Organization of Study   5
   - Notes   6

2. **AN INDUSTRY RESTRUCTURED**   7
   - Overview of Changes   7
   - The Traditional U.S. Machine Tool Industry   8
   - Technological Trends Shaping the Industry   18
   - Economic Trends   23
   - New Entrants and New Competitive Strategies   32
   - Response of Machine Tool Builders to These Changes   38
   - Conclusions   45
   - Notes   48

3. **THE DEPARTMENT OF DEFENSE, PRIME CONTRACTORS, AND THE MACHINE TOOL INDUSTRY: RELATIONSHIPS THAT AFFECT INDUSTRY STRUCTURE**   52
   - Size of DOD and Contractor Markets   53
   - DOD Procurement: Incentives and Disincentives   55
   - Manufacturing Technology Programs   56
   - Machine Tool Suppliers' Perspective on the Defense Procurement Process   63
   - Perceptions of the U.S. Machine Tool Industry: The Prime Contractors' Viewpoint   69
   - Domestic Legislation Affecting the Purchase of U.S. Produced Machine Tools   76
   - Conclusions   79
   - Notes   81

4. **PROBLEM SYNTHESIS AND RECOMMENDATIONS**   84
   - Problem Synthesis   84

	Recommendations	88
	Conclusions	96
	APPENDICES	99
A.	Highlights of Phase I Study	101
B.	Policies of Foreign Governments	111
C.	Questionnaire Sent to Machine Tool Industry Executives	118

1

INTRODUCTION

STATEMENT OF THE PROBLEM

The performance of the U.S. machine tool industry has a major impact on the efficiency, effectiveness, and timely production of defense materiel, despite its relatively small share of the national economy. To provide for the national security, the Department of Defense (DOD) manufactures and procures a wide variety of articles, which depend in turn on a wide variety of manufacturing processes. To carry out this mission effectively, DOD needs not only materials but continuing access to the latest process technology to cut and shape those materials into required components. In addition, the DOD mission needs expandable capacity to manufacture both finished articles and spare parts during mobilization and extended military conflict.

Recent trends, including a sharp surge in machine tool imports as a percentage of domestic consumption, have called into question the ability of the domestic machine tool industry to meet current needs for defense production under both peace and wartime conditions. The Department of Defense requested the formation of this Committee to assess the international competitiveness of the domestic machine tool industry, study its current and expected responsiveness to defense needs, and recommend actions and policies for DOD and others to ensure access to a sufficient machine tool capacity and capability.

THE U.S. MACHINE TOOL INDUSTRY: THE PROBLEMS OF MATURITY

The U.S. machine tool industry shows many of the characteristics of an aging, mature industry. Annual growth of real domestic machine tool output has stood at approximately 0.1 percent for the last decade; average annual productivity, measured by output per man hour, actually declined during 1973–1981. Contributing to the industry's low productivity is the fact that its own production machinery is relatively old. in 1978, 40 percent of its machines in use were over 20 years old.[1] In Japan, by contrast, the comparable figure has been estimated at 18 percent.[2]

Like other mature industries such as steel, the U.S. machine tool industry has been hit hard by foreign competition. Machine tool imports, which stood at 9.7 percent of domestic consumption in 1973, climbed to 24.2 percent in 1981.[3]

Adding to the problems of the domestic machine tool industry are some far-reaching technological advances that not only are altering the types of machines being demanded by end users but have also given rise to the entry of new types of firms in the provision of machine tools in the broader market for factory automation products. In various stages of research, development, and implementation are (1) synthetic materials, such as composites, ceramics, and plastics, that will ultimately replace metals in some military and civilian applications; and (2) new processes for forming and working both metals and other materials, which will reduce the need for traditional machining. In addition, the growth of computer-integrated manufacturing has meant that new sets of firms (e.g., manufacturers of computer controls) are entering the broader market. While none of these firms have entered as manufacturers of machine tools per se, they—along with specialized assembly firms and machine tool builders themselves—are likely to become major players in the process of fitting machine tools with computer technology. Accordingly, U.S. machine tool builders will have to adapt to new markets and new products.

To remain competitive under these conditions will require (1) massive investments by the U.S. machine tool industry in research and development; (2) a substantial broadening in these companies' R&D, engineering, and software capabilities; (3) a reshaping of their development strategies; and (4) heavy investment in modern production facilities. It is uncertain, however, how many of the companies in the U.S. machine tool industry other than the industry's leaders have the ability or the perception of necessity to accomplish these tasks, although some individual firms representing a significant part of the industry's production are already engaged in meeting the challenge.

This report concludes that in the face of urgent competitive pressures, some U.S. machine tool builders have already begun responding to the challenges of new competition and technology. As the Phase I study indicated, the machine tool industry as traditionally defined is giving way to a more sophisticated one, which is also engaged in, for example, factory automation and computerized controls. The Committee believes that, given a sustained economic recovery and aggressive steps by both government and industry, an effectively competitive domestic machine tool industry can emerge. This industry, however, will be substantially different from the machine tool industry as traditionally defined; many traditional firms who are unable or unwilling to take the appropriate steps to modernize will not survive the rapid changes that are now upon them. Without such a transition, the United States may lose or seriously damage a resource that is valuable to the national economy and the national defense. Indeed, one of the aims of this report is to describe what the "survivors" of this transitional phase in the machine tool industry will look like and how they will get there.

THE DOD INTEREST

As this report describes, the Department of Defense already manages several programs aimed at improving manufacturing productivity and maintaining a reserve of machine tools. The Committee found that the DOD's interest regarding the U.S. machine tool industry includes having access to state-of-the-art technology; being able to utilize cost-effective, expandable production facilities; and having a macro-economy that permits long-term growth within the domestic machine tool industry.

The changes in technology and markets referred to above, however, suggest that DOD's interest is tied to the international competitiveness of a "restructured" machine tool industry substantially more complex than the traditional industry as defined today. This scenario necessarily places more emphasis on measures to modernize the industry, and less emphasis on such traditional measures as stockpiling maintenance.

The changing status of the machine tool industry raises difficult questions over how the government should treat mature, basic industries that are beset by rapid change. In many such industries, conventional business economics has seemed to favor the offshore manufacturing facility, whether this facility be owned by a U.S. or a foreign company. This tendency has recently appeared with respect to machine tools. Thus approximately 40 U.S. machine tool firms have overseas facilities. On the other hand, some foreign manufacturers (e.g., Yamazaki, Hitachi-Seiki, Oerlikon-Motch) have established manufacturing or assembly plants in this country.

The task facing this Committee, therefore, has been twofold: first, to collect the data necessary to draw valid conclusions as to the health and future of the U.S. machine tool industry; and, second, to assess the implications for the national security of these conclusions.

The Committee believes that certain policy changes are vitally important in support of this transition. Chapter 4 of this report contains recommendations for action by DOD, other government agencies, prime contractors, and machine tool builders themselves.

APPROACH TO THE STUDY

As the second of two studies produced for the Department of Defense on the subject of the machine tool industry's international competitiveness (see Preface), the work of this Committee stems from the work of the Phase I committee and its report.

During the research phase, the Committee conducted written surveys, site visits, and interviews. Two written surveys were conducted: one of machine tool users that had made recent purchases, to learn what they had bought and why; and one of machine tool builders, to learn their perceptions of recent economic and technological trends.

Eleven site visits were conducted at firms chosen to represent a wide spectrum of machine tool firms. The major groups having an interest in this study—DOD, machine tool builders, prime contractors, and major subcontractors—participated in a total of several dozen telephone interviews, in addition to the surveys and site visits.

These primary data collection efforts were augmented by a literature review and the collective knowledge of the Committee members. When the data gathering effort was completed, the Committee undertook to synthesize the views presented by the various sources.

From the start, the Committee was impressed with the unevenness of the data. Much of the available information is highly aggregated, thus obscuring the situation of many individual firms, or merely anecdotal, thus making generalizations difficult. Under these circumstances, the Committee was forced to rely in a number of instances upon its own surveys, as well as the subjective judgments of its members.

The Committee broke the issues down into the following questions, to form the underpinnings of its analysis.

- What is the technological and economic state of the U.S. machine tool industry today relative to foreign competition?
- Is the U.S. in danger of losing two important strategic resources: its machine tool manufacturing capability and its position as a leader in manufacturing process technology?
- What are the causes of the problem of increased import competition in the machine tool industry?
- To what extent has DOD action affected the current status of the U.S. machine tool industry?
- Are there major shortcomings in the machine tool industry structure and performance that are in the national interest to change?
- What are the national security interests regarding the U.S. machine tool industry?
- What constructive contributions might be provided by DOD in pursuing these interests?
- What are the potential contributions of other executive branch government agencies, prime contractors, and the U.S. Congress?
- What policies and actions should be applied by the machine tool industry?

ORGANIZATION OF STUDY

The Committee organized its written analysis according to two broad topics: (1) the present competitive situation of the U.S. machine tool industry; and (2) the relationships among DOD, prime contractors, major subcontractors, and machine tool builders that affect the competitiveness of the U.S. builders. These are described in Chapters 2 and 3, respectively. Chapter 4 presents the Committee's conclusions about the implications of this situation for this country's national security goals, and presents a set of recommended options for DOD and others. Three appendices are also included.

NOTES

1. National Machine Tool Builders' Association (NMTBA), Economic Handbook 1982/83, p. 233.

2. Japan Productivity Center. More recent (1981) data suggest that while the total inventory of Japanese machine tools is not as young as it once was, it is still higher than the U.S. for most machine tool categories. Anderson Ashburn, "Modernization Pace Slows in Japan," America Machinist (January, 1983), p. 122.

3. NMTBA, Economic Handbook, 1982/83, p. 126.

2

AN INDUSTRY RESTRUCTURED

OVERVIEW OF CHANGES

We're going through a <u>revolution</u> in manufacturing technology. Formerly, you would have talked about <u>evolution</u>. [Director of manufacturing research at helicopter plant]

The American machine tool manufacturer is not as competitive as his [foreign] competitors are...not as prepared to make changeovers into new technology. [Head of facilities division at aerospace firm]

The U.S. machine tool industry is undergoing fundamental restructuring. A structurally more complex and technologically dynamic industry is replacing a mature, less complex one.

The industry has been characterized by fragmentation, relatively low levels of capital investment, and conservative management. Strong forces from outside the domestic machine tool industry, however, have made this traditional posture of the industry permanently outmoded. These forces include technological as well as economic factors: for example, the increasing use of new technologies in machine tool construction and applications, and the increasingly global view of machine tool markets by foreign suppliers.

The machine tool industry has undergone fundamental change over the past decade. Although basic metal-cutting and metal-forming machines are still a critical element in the manufacturing picture, the machine tool industry today is becoming part of a new, automated manufacturing industry that is producing new types of products, such as computer-driven, integrated production systems, that did not exist 10 years ago. It contains new industry segments which have entered the market to promote advanced technologies. It is diversifying into the processing of new materials. And it is today more than ever part of a world market, with worldwide sources being used even by U.S. machine tool firms. In this world market, however, the U.S. firms are being seriously challenged by foreign manufacturers instead of dominating markets as they did 10 years ago. It is this new, broader, and worldwide industry that forms the basis for an assessment of the machine tool industry's responsiveness to national security needs.

This chapter traces how these developments are restructuring the U.S. machine tool industry today, and is divided into the following sections:

- the traditional U.S. machine tool industry
- technological trends shaping the industry
- economic trends
- new entrants and new competitive strategies
- response of machine tool builders to these changes

THE TRADITIONAL U.S. MACHINE TOOL INDUSTRY

Any analysis of the machine tool industry in the United States today must incorporate the fact that technological and market conditions are altering the definition of the industry and the players in it. To impart some appreciation of these changes, this report starts with an examination of the traditional U.S. machine tool industry.

<u>Definition</u>. According to the National Machine Tool Builders' Association (NMTBA), the industry comprises the manufacturers and sellers of machine tools, defined as "power-driven machines, not hand held, that are used to cut, form or shape metal."[1] Metal-cutting machines include lathes, grinding machines, milling machines, and machining centers. Typical metal-forming machines are presses, forges, and punching, shearing, and bending machines. This product classification conforms to the Standard Industrial Classification Codes 3541 (metal-cutting) and 3542 (metal-forming).

<u>Size</u>. The machine tool industry, thus defined, is a relatively small sector of the economy. Production in the United States totaled $3.6 billion in 1982, or 0.12 percent of GNP.[2] Total employment in the industry at the end of 1982 was estimated at 68,000, or less than 0.10 percent of U.S. employment.[3]

Until 1971, when U.S. machine tool production was outstripped by West Germany, the United States had been the world's leading producer of machine tools since the end of World War II. The United States regained the lead in machine tool production in 1979, only to be surpassed by Japan in 1982.[4] The growth of Japanese machine tool production has been especially swift, averaging approximately 30 percent annually between 1976 and 1981.[5]

Notwithstanding this development, American machine tool builders have sold, and continue to sell, many machine tools for export. Exports have averaged 13.8 percent of domestic machine tool production during 1971–1981, and in 1981 itself stood at 13 percent of domestic production. These exports have also held their own compared with the combined exports of other countries. The U.S. share of world machine tool exports has ranged about 8 to 10 percent since 1970; the rapid rise in Japanese machine tool export trade, in fact, appears to have been more at the expense of West German than of U.S. exports.[6]

The Committee notes in passing that the Eastern Bloc countries, once a large market for U.S.-made machine tools, have now effectively disappeared as significant purchasers of U.S. equipment. Machine tool exports to these countries stood at $92.5 million in 1975, or 16.3 percent of total such exports; the corresponding figures for 1981 are $22.8 million, amounting to 2.2 percent of exports. Machine tool exporters, therefore, have had to find other markets to compensate for this loss. Although the Committee's mandate did not include pursuing this issue further, the Eastern Bloc sales situation is viewed by the Committee as an "unsolved" question that merits further U.S. government attention.

Concentration. Most companies in the U.S. machine tool industry have traditionally been small, closely held firms with narrow product lines. Table 1 shows the extent to which small establishments have populated the industry.

In addition, the industry has not been characterized by significant firm concentration. According to Commerce Department figures, the 4 largest metal-cutting machine tool establishments were responsible for 22 percent of industry shipments in 1977. In 1981, 15 companies accounted for approximately 70 percent of the machine tool industry's shipments, as Table 2 shows. This means that the other 30 percent of shipments came from the remaining 1,000-plus establishments.

TABLE 1 Size of U.S. Machine Tool Establishments

	1963	1967	1972	1977
Number of establishments	1,167	1,253	1,277	1,343
Average size of establishment (employees)	71	93	60	62
Percent with 20 or more employees	36	40	34	35

Source: NMTBA, <u>1982–83 Economic Handbook of the Machine Tool Industry</u>.

Sales Pattern. Machine tool sales have traditionally been sensitive to changes in the business cycle. The National Academy of Engineering recently observed that "perhaps the most important trait associated with the machine tool industry is the extreme cyclicality of its income, profits and cash flow." It concluded that "it would be impossible to understand the American machine tool industry without appreciating both the depth and wide-ranging implications of these cycles."[7] Year-to-year swings in machine tool orders of +75 percent and −50 percent have occurred (see Figure 1),[8] compared with maximum sales swings of +32 percent and −34 percent in steel.[9] This sales pattern has forced upon the industry a strategy of "buffering" business cycle downturns by accumulating order backlogs from boom times.[10] As the following paragraphs indicate, this pattern has prevented even large machine tool firms from having the capital investment, R&D, and overseas sales structure found in other manufacturing firms (e.g., office equipment) of similar size.

Employment Patterns. This cyclicality has had an effect on employment in the industry. Although the industry generally pays its employees better than the average of durable goods manufacturers (see Table 3),

AN INDUSTRY RESTRUCTURED

whether because of differences in skill levels or employment conditions, employment fluctuations have been substantially sharper among machine tool companies than in the durable goods sector of the nation as a whole. Commerce Department figures show that average changes in machine tool production worker employment are more than one and one half times the percentage changes in durable goods employment generally.

TABLE 2 Shipments by the 15 Largest U.S. Machine Tool Companies

Company	Estimated 1981 Shipments of U.S.-produced Machine Tools ($millions)
Cincinnati Milacron	498.0
Bendix	400.0
Cross & Trecker	310.0
Giddings & Lewis	286.9
Ex-Cell-0	280.0
F.Joseph Lamb	275.0
Textron	270.0
Acme Cleveland	240.0
Litton	200.0
Ingersoll Milling	200.0
White Consolidated	180.0
Gleason Works	160.0
Houdaille	150.0
Monarch	140.1
Esterline	112.3
	3,702.3=73% of
Total Shipments 5,095.6	

Sources: American Machinist, August 1982, p. 51; NMTBA.

Industry observers, and the Committee's own surveys, cite this cyclicality as one of the causes for the industry's conservative management and the inability of many machine tool firms to attract and retain the brightest engineering, managerial, and technical talent.

Profitability. A common assertion has been that machine tool industry profitability is somewhat higher than the manufacturing average during upturns in the business cycle, but substantially lower on the downside.[11] Table 4 sets forth financial ratios that contradict this general assertion at least for the years 1975–81. These ratios indicate that the industry has maintained moderately healthy levels of profits and earnings relative to sales and to net worth, that these levels have risen since the middle of the last decade, and that they compare favorably with corresponding ratios for durable goods manufacturers. In 1982 and 1983, however, many U.S. machine tool companies posted significant losses[12] and at least one prominent industry analyst has commented that "the machine tool industry faces difficult profitability through 1984."[13]

FIGURE 1 Year-to-Year Change in Real Net New Orders of Machine Tools, 1957–82.

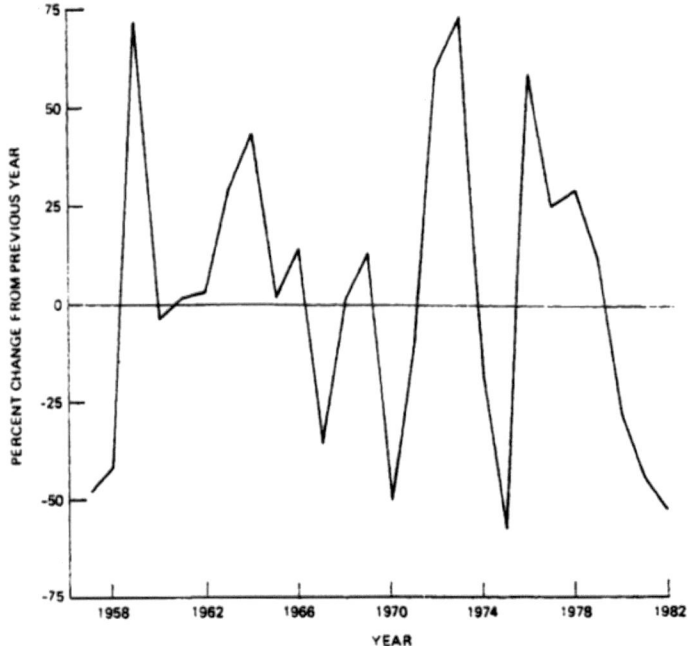

Sources: NMTBA, Economic Handbook of the Machine Tool Industry, 1982–83; NMTBA, "Industry Estimate of New Orders, Cancellations, Shipments and Backlog (monthly)"

TABLE 3 Wage Rates for Metal-Cutting Machine Tool Employees Relative to Durable Goods Hourly Wages

Year	Percent
1960	106
1965	110
1970	112
1975	108
1976	108
1977	109
1978	109
1979	109
1980	109
1981	107
1982	108

Source: U.S. Department of Commerce.

Research and Development. Conventional machine tool industry managers have been cited by outside observers[14] and by members of the machine tool industry itself[15] for taking a short-term perspective on their market. Technological pre-eminence and a reputation for excellence are difficult to maintain without investment in basic research and development. The willingness and ability to invest in R&D requires a long-term outlook and an understanding that state-of-the-art technology and its potential for developing new products are essential for survival.

The Committee found that a few leading machine tool companies have maintained R&D initiatives. However, the industry as a whole has traditionally drawn on outside sources for new technology and new product development—e.g., from the manufacturers of computers and controllers, manufacturing systems designers, and DOD prime contractors—rather than from internal R&D efforts. As this report points out, this pattern of technology flow can be ascribed in part to conventional defense procurement practices.

TABLE 4 Selected Financial Ratios Comparing U.S. Machine Tool Industry with Durable Goods Manufacturers

	Machine Tool Industry		Durable Goods Industry		Comparisons	
	a.	b.	c.	d.	e.	f.
Year	Net Operating Profit on Sales	Earnings on Net Worth After Taxes	Net Operting Profit on Sale	Earnings on Net Worth After Taxes	column "a" as % of column "c"	column "b" as % of column "d"
1975	9.1	13.8	6.7	9.9	135.8	139.3
1976	9.4	11.0	7.9	13.6	118.9	30.8
1977	7.6	12.3	8.2	14.5	92.6	84.8
1978	7.8	12.8	8.5	15.9	91.7	80.5
1979	12.2	16.3	7.6	15.5	160.5	105.1
1980	13.1	18.1	6.0	11.2	218.3	161.6
1981	12.6	18.0	6.5	12.0	193.8	150.0
1975–81 averages	10.25	14.61	7.34	13.22	144.51	114.58

Source: Federal Trade Commission; NMTBA

Data on R&D outlays by the U.S. machine tool industry are contradictory. Two independent sources estimate that R&D investment averaged 1.5 to 1.6 percent of sales over the past decade.[16] Figures supplied by the industry on a confidential basis to their trade association put the level at 4.1 percent.[17] The NMTBA's data report that R&D climbed to 4.2 percent of sales in 1981 and 1982, reflecting either new R&D initiatives and/or the industry's inability to cut R&D below certain minimum levels during recessionary periods.

Analysis of this issue is complicated by the fact that the definition of "research and development" in the machine tool industry is not uniform. Because much of the industry's work involves the adaptation of basic machine tools and manufacturing systems to specific customer requirements, many machine tool companies include such engineering application expenses with their R&D accounts. As a result of this accounting practice, which is not unique to the industry, machine tool industry R&D ratios may be inflated.

The dollar amounts spent on R&D in the domestic machine tool industry also shed some light on that industry's economic situation. Table 5 sets forth these amounts, on both a current and a constant dollar basis.

What the table shows is that R&D outlays have been heavily affected by economic slowdowns, and in 1982 fell almost to the level of outlays, in real terms, that existed in 1975.

TABLE 5 Research and Development Outlays, U.S. Machine Tool Industry

Year	Current Dollars	Constant (1975) Dollars
1975	$ 73,174	$ 73,174
1976	73,231	67,175
1977	83,238	70,541
1978	104,855	79,436
1979	128,216	84,911
1980	171,539	96,915
1981	188,196	96,018
1982 (est.)	151,385	75,693

Source: NMTBA.

Capital Investment. Table 6 compares capital outlays in the machine tool industry (SIC Codes 3541 and 3542) with outlays in related industrial sectors. It shows that U.S. machine tool industry outlays for capital spending have generally lagged those of other industries. This is consistent with the conclusion, referred to above, that machine tool builders have tended to rely on stretched out order backlog management, rather than increased capacity, to accommodate cyclical changes in demand.

Growth and Productivity. The U.S. machine tool industry's share in world machine tool production is significantly below what it was in the late 1960s. In 1968, for example, the U.S. share in world machine tool output was more than 25 percent. Since 1970, however, it has failed to climb above 20 percent.[18]

Of more significance, because of its implications for the future, productivity growth in the U.S. machine tool industry has also been poor. Table 7 compares machine tool industry output and productivity growth with cor responding figures for the U.S. durable goods sector. It shows that machine tool industry productivity growth has averaged a negative 0.7 percent annually during 1973–1981, which is substantially less than the performance of U.S. durable goods industries during the same period.

TABLE 6 New Capital Expenditures as a Percent of the Value of Shipments—Selected Industries, 1975–1980

Industry	Percent
Miscellaneous Machinery (SIC 359)	5.9
Office Machinery (SIC 357)	5.5
Blast Furnaces/Basic Steel Products (SIC 331)	4.6
Construction Machinery (SIC 353)	4.0
General Industrial Machinery (SIC 356)	3.5
Engines and Turbines (SIC 351)	3.4
Motor Vehicles and Equipment (SIC 371)	3.4
Farm Machinery (SIC 352)	3.1
MACHINE TOOLS (SIC 3541 AND 3542)	2.9
Special Industrial Machinery (SIC 355)	2.9
Refrigeration and Service Machinery (SIC 358)	2.5

Sources: Based on data from the Annual Survey of Manufactures and the 1977 Census of Manufacturers.

Although it is possible that some productivity loss could have been caused by the retention of skilled workers during economic downturns, the majority of the Committee believed that the productivity growth record bears some relation to the levels of capital investment and R&D within the industry. While the connection cannot always be measured directly, it is generally accepted that high levels of capital investment and R&D spending are essential to maintaining productivity growth in technology-intensive industries.[19]

Marketing. This general picture of a not very robust domestic industry is also reflected in the marketing practices of U.S machine tool builders. The industry itself has recognized that machine tool company management needs to adopt a long-term outlook and willingness to invest in effective marketing networks that its Japanese competitors have.[20] Interviews conducted for this report, however, revealed that marketing strategy for U.S. machine tool firms is usually reactive and has tended to concentrate almost exclusively upon the stated needs of its larger, U.S.-based customers, with little development of a more varied customer base. Japanese marketing efforts in the United States, on the other hand, began with a focus on mass-produced, low-unit-cost numercial control (NC) machine tools attractive to small and medium-sized users. A further description of Japanese machine tool marketing efforts in the United States is given later in this chapter.

TABLE 7 Growth of Output and Productivity: Annual Average Percent Change

	1959--1973	1973--1981
Growth of output	4.6	2.3
Manufacturing	4.8	2.5
Durable goods manufacturing	2.3	0.1
Machine tool industry	2.3	0.1
Growth of output per hour of all employees		
Manufacturing	3.0	1.7
Durable goods manufacturing	2.8	1.7
Machine tool industry	1.0	−0.7

Source: Bureau of Labor Statistics, U.S. Department of Labor.

Concluding Comments on the Traditional U.S Machine Tool Industry. The above paragraphs describe an industry that has lost its position as the world's number one producer of machine tools, to a nation whose own machine tool industry has been experiencing dramatic growth which does not appear to be slowing. This decline in the U.S. industry's position fits the pattern of other mature, domestic industries that have in the course of a small number of years come under severe competitive pressure from younger, foreign-based firms. In the case of the U.S. machine tool industry, this pattern has evidently been accentuated both by industry structure (e.g., fragmentation) and by the practices of industry management (e.g., failure to adopt a global, longer-term view of markets). This structure and these practices have influenced decisions regarding capital investment, R&D, marketing, and employment. These decisions seem to have left the industry ill-equipped for necessary large investments in new technology and new marketing efforts: factors which, as the following sections show, are key to the industry's future.

TECHNOLOGICAL TRENDS SHAPING THE INDUSTRY

In its 1982 annual report, Cincinnati Milacron stated that 49 percent of its sales were of products that it did not make five years ago.[21] This observation, coming from one of the most forward-looking U.S. machine tool firms, demonstrates the challenge facing the entire industry. The key technological trends giving rise to these new products are (1) the increasing use of computers in factory automation; (2) the increasing use of substitute materials, some with applications that permit substitution for metals; and (3) new methods for metals processing.

Computers and Automation

Approximately 10 years ago, most machine tools sold were manually operated, stand-alone machines. Today, such machines remain economically appropriate for many applications, but in 1982, 36 percent of the machine tools purchased in the United States were operated by "numerical" (usually computer) control rather than manually.[22] Automation of a machine tool's function via numerical control (NC) has been available to manufacturers for almost three decades. Moreover, higher levels of automation which incorporate not only an individual machine's function but the material handling and control systems as well, are already in use today in metal fabrication and are likely to become commercially more attractive in the near future. Indeed, the new, rapid pace of automation driven by the need to build flexibility into, and costs out of, manufacturing operations is giving rise to forecasts by industry experts of a boom market for automated factory equipment over the next decade.

By the end of this decade, flexible manufacturing cells and systems are likely to be in high demand. Flexible manufacturing systems (FMS) represent a new application for machine tools in which groups of machine tools are integrated and controlled by a central computer; the same computer also controls integrated materials handling systems, including robots, that move workpieces from machine to machine and position a workpiece at each machine.

Broadly speaking, then, a new manufacturing process industry is developing whose products will soon include not only machine tools as traditionally defined and computerized controls for individual machine tools, but also more complex computer hardware and software, materials handling systems, machines for assembly, testing, washing, plating, and heat-treating components; and robots. This phenomenon is driven, if by no other reason, by events taking place around the globe. The Japanese Study Mission of the National Machine Tool Builders' Association, for example, reported in 1981 that throughout its travels in Japan,

> it was apparent that FMS is upon us. Virtually every Japanese [machine tool] builder was talking about it, preparing products for it and planning to use it in his plants. Several builders have manufactured and sold FMS systems and at least two of them have complete FMS-equipped parts-making plants under construction.[23]

This observation illustrates the increasing emphasis on <u>integration</u> in the machine tool industry, wherein traditional machine tools are used as parts of larger manufacturing systems incorporating the products of non-machine tool manufacturers such as computer-makers.

An even greater degree of automation than individual flexible manufacturing systems could become commercially attractive in a robust economy, and a necessity in view of the heavy pressures of international competition for lowering production costs. Turnkey automated factories have been designed for industries, such as chemicals, cigarettes, paper, and textiles, that do not use metalworking machine tools. Among the prospects for automated factories in the U.S. metalworking industry are plants composed of flexible manufacturing systems where inventory management, scheduling, and routing are all computer controlled, and where robots and other automated equipment for such non-metalworking tasks as painting are also controlled by the same central computer that coordinates metalworking production. In addition, for both FMS and automated factories, the specifications of the parts to be produced can be developed by computer-aided design (CAD) equipment that will be connected to the rest of the production system with the central computer.

Increasingly, therefore, new products will arise from the integration of information, electronic, and mechanical technologies. Figure 2 attempts to provide a pictorial representation of how technology is changing the face of the machine tool industry.

Although there is substantial evidence that the U.S industry is at least as technologically advanced as the Japanese in the technology of FMS and other factory automation,[24] the chief difference between the machine tool industries of both countries seems to be in the <u>application</u> of that technology. As the NMTBA's Japanese study mission further observes,

> a new parts-manufacturing plant in America would, most probably, be a modernized version of existing plants. In Japan new machine tool parts making plants use only the Latest technology.[25]

In all countries, the market for complete flexible manufacturing systems still remains small relative to the economy as a whole. The importance of the FMS concept, however, goes beyond the number and growth of complete systems. More modest, partial, or limited applications of entire automated systems are widespread, consisting of production subsystems and incremental stages of planned FMS projects.

From the seller's standpoint, therefore, the market for FMS components is considerably larger than the number of companies capable of purchasing an automated factory. Of the estimated 1,300 or more machine tool establishments in the United States, approximately 37 companies (the count will vary according to the way subsidiaries and divisions of firms are accounted for) claim to be able to manufacture complete manufacturing systems. Six of those firms are among the fifteen largest machine tool firms cited in Table 2, above; together those six firms account for approximately $1.87 billion in machine tool sales. A larger number of firms manufacture equipment, such as computerized controls, programmable robots, CNC controls, and materials handling devices, that is ancillary to the machine tools in such systems. Significantly, a number

FIGURE 2 Technology Map

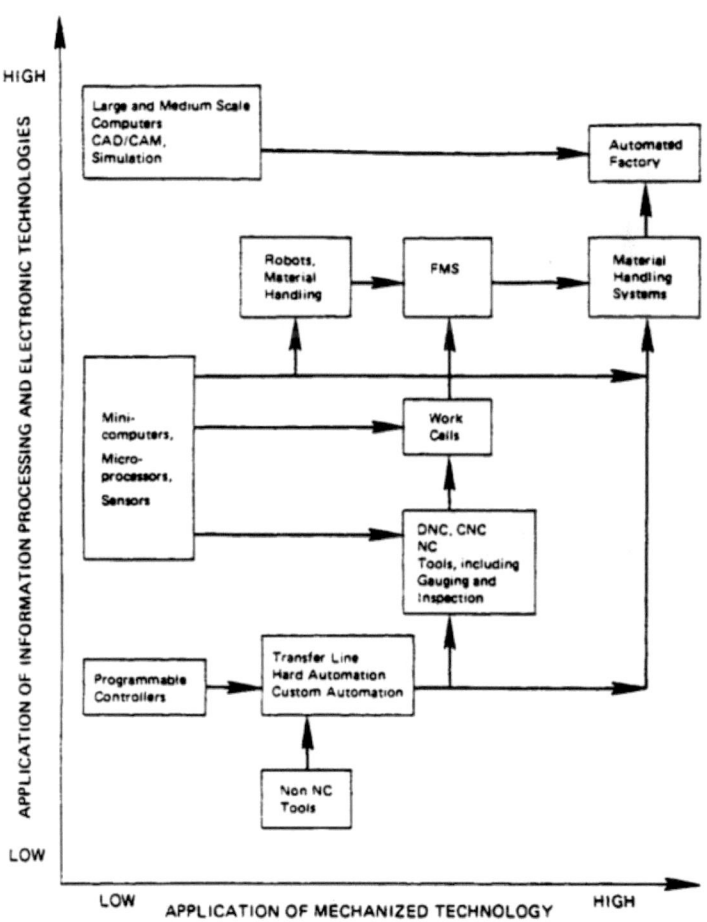

of companies other than traditional machine tool companies are also beginning to participate in this marketplace.

New Materials and Methods

Another development confronting traditional machine tool companies is the potential for reduction of metal-cutting and metal-forming markets through (1) the displacement of conventional metals by new materials such as composites, powdered metal, ceramics, and plastics; and (2) the introduction of new techniques that reduce the amount of machining required to produce a finished shape.

Non-metal Materials. At present, metals have not been supplanted by other materials to a degree that has significantly affected machine tool markets. Nevertheless, depending upon their application, non-metals can be less expensive to manufacture; can be produced to net or near net shape; have superior performance because of greater strength, less weight, and durability; and save energy.

Composites have been used in airframe construction since the late 1960s, and applications have grown as manufacturers have gained more experience and confidence in composite technology. For example, the P-14 and P-15 aircraft have a relatively small amount of composite material, approximately 3 percent of structural weight. The newer F-18 and AV8B fighters contain 13 and 26 percent composites, respectively. These percentages should continue to grow as new airframes are developed. In other words, each new generation of aircraft is likely to have a substantially greater proportion of composites. There are conflicting indications as to how rapidly significant applications may occur, however, particularly in large airframes, due to long development lead times for new aircraft.[26]

These new materials represent a potential growth area for the machine tool industry. For example, plastics shipments already exceed the tonnage of steel, aluminum, or copper shipments.[27] This large volume has created a potential growth market in plastics processing machines. According to the 1983 National Machine Tool Builders' Association Directory, eight NMTBA members manufacture plastics forming equipment.[28] One of these firms (Cincinnati Milacron) is among the largest 15 U.S. machine tool companies.

Metal Processing. New technology for metal processing can, in special circumstances, reduce the need for conven tional metal finishing, and also increase the precision of the metal fabrication process. Advances in near-net-shape forming techniques via investment casting, powder metallurgy, and continuous extrusion all produce metal parts close to their final form, with savings in material scrap and machining costs. New metal-cutting and -removing technology includes lasers and chemical milling. Recent developments in metal-forming and -surfacing technology include electro-deposition and ion implantation. Commercially, none of these technologies has substantially displaced prevailing metal-cutting and -forming technologies, and this report does not undertake to estimate the pace of development of these technologies or the scope of their ultimate military and commercial application.

At the present time, approximately 10 U.S. machine tool firms are involved in these new technologies.[29] Five of the fifteen largest U.S. machine tool firms, with combined machine tool sales of approximately $900 million, are among these ten.

The Committee believes that technology flows will continue to shape the industry. The machine tool business worldwide has become a fast-moving sector, technologically, where the United States cannot afford to be outdistanced by countries whose machine tool manufacturers take a more aggressive approach to pushing technological advances within their own firms.

ECONOMIC TRENDS

The U.S. machine tool industry has been substantially influenced not only by technological trends but also by economic ones. The recent recession, which has been the steepest of the eight postwar recessions, and the loss of sales to foreign producers have cut seriously into profits. It produced losses in some cases even as the Administration has put into place an expanded defense budget, and even as the signs are increasingly evident that a recovery is under way. New machine tool orders are considered a "lagging" economic indicator; also, an indicator, that fluctuates more widely than do other series such as industrial production. This means that, although machine tool order levels are recovering somewhat as expected, it could be a year or more before orders reach levels that signal a recovery in the machine tool sector itself, and much longer before earnings can

Globalization of Machine Tool Competition

The globalization of direct machine tool competition is perhaps the most significant economic trend in the domestic industry today, for it is a new and permanent one.

International trade in machine tools is not new. In 1949–51, for example, approximately 20 percent of U.S. machine tool production was sold abroad.[30] What is new is that sophisticated machine tool building industries have now developed since then in a number of nations, and a substantial number of these foreign machine tool firms are able to compete globally.

In addition, the proportion of world trade in machine tools has grown. In 1968, 29 percent of world machine tool output was exported; the corresponding figure for 1981 is 40 percent.[31] As part of this trend, the propensity for global sources in the machine tool industry (i.e., looking beyond national boundaries for machine tools and components) is becoming more pronounced. As this report points out, the U.S. machine tool industry itself, by locating more of its own manufacturing facilities overseas, is participating in this trend toward global sources. Although U.S. machine tool firms have consistently been strong exporters, foreign firms are capturing increasingly large shares of the domestic U.S. market.

Figure 3, which traces the U.S. trade balance in machine tools, shows that in 1978 the United States for the first time imported more machine tools than it sold abroad. This trade imbalance in machine tools has increased since then. In 1981, the U.S. trade deficit in machine tools was $482 million. As shown in Table 8, imports as a percent of U.S. consumption have risen very rapidly, from less than 10 percent in 1973 to 24.9 percent in 1981 and an estimated 26.8 percent in 1982. The NMTBA estimate for the first quarter of 1983 is 33.8 percent.

One way of comprehending the effect that the acceleration in imports has had on the sales of domestic machine tool firms is to hypothesize what might have happened if imports had grown simply at the same rate as U.S. machine tool consumption. The Committee has calculated that had imports of Japanese machine tools since 1975 grown at the same rate as U.S. machine tool consumption during the same period, the Japanese market would be approximately $500 million less today than it is. This $500 million figure is equivalent to the sales of the largest U.S. machine tool firm and corresponds to approximately 10.2 percent of shipments of U.S.-made machine tools.

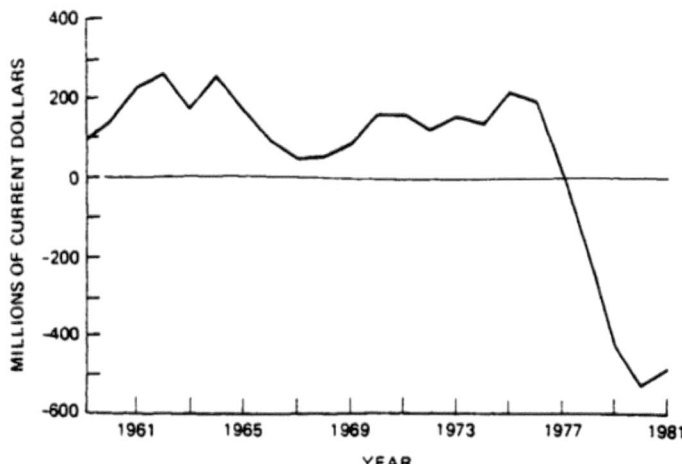

FIGURE 3 U.S. Trade Balance in Machine Tools. Source: NMTBA, Economic Handbook of the Machine Tool Industry 1982–83, p. 123.

The country with the most rapid growth in machine tool trade, and in the value of machine tool exports to the United States, has been Japan. Japanese machine tool exports to the United States surged from $22.1 million in 1973 to $687.5 million in 1981. Although Japanese machine tools were less than 15 percent of total U.S. machine tool imports in 1973, they accounted for nearly half of such imports in 1981. Imports from Western Europe, measured as a percent of domestic consumption, have remained generally constant.

It is worth examining more closely the elements behind the initial Japanese success in the U.S. machine tool markets, because they bear some relation to the difficulties found by the U.S. industry. In its interviews and deliberations, the Committee found five such elements that deserve mention: delivery times, reliability, targeting, prices, and commercial and government policies and practices.

TABLE 8 U.S. Machine Tool Imports and Exports

	1973	1976	1979	1981	1982
U.S. Machine Tool Production					
(millions of dollars)	$1,788.9	$2,178.3	$4,064.0	$5,111.3	$3,744.0
U.S. Machine Tool Exports					
Amount (millions of dollars)	224.7	352.3	395.6	671.6	426.6
Percent of production	13.1	16.2	9.7	13.1	11.4
U.S. Machine Tool Imports					
Amount (in millions of dollars)	167.1	318.3	1,043.8	1,217.0	
Percent of consumption					
All machine tools	9.7	14.9	22.2	24.9	26.8
Metal-cutting types	10.7	15.6	23.2	25.4	n.a.
Lathes (excl. vertical turret lathes)	15.0	18.6	39.5	55.6	n.a
Amount of imports (millions of dollars) from:					
Japan	22.1	67.2	352.8	687.1	n.a
West Germany	51.6	93.3	197.3	191.8	n.a
Percent of imports from:					
Japan	13.2	21.1	33.8	48.0	n.a
West Germany	30.9	29.3	18.9	13.4	n.a

Source: NMTBA, Economic Handbook 1982–83.

Delivery Times. The traditional practice of order backlog management, which served U.S. machine tool builders well for several decades, was based on an implicit assumption that potential foreign competitors did not have the resources to take advantage of wide swings in the U.S. machine tool market. Whether this assumption was ever valid, it certainly was not so by the late 1970s. By that time, many foreign firms had the resources to offer fast delivery of quality machines to U.S. customers who did not wish to wait for backlogs to be worked down by their domestic suppliers.

Figure 4 compares imports of machine tools with unfilled domestic orders. It shows a relatively close correlation between surges in the backlogs, and in imports. The only major break has come in the last three years, when imports have continued to rise (albeit at a slower rate) while unfilled orders fell because of the recession.

The figure confirms the Committee's judgment that one important reason for Japan's success in the U.S. market has been this delivery time factor. Surveys conducted for this report reveal that U.S. manufacturers were able to obtain delivery of Japanese machines within one or two months during the late 1970s, when some domestic builders were requiring a 1–1/2 to 2 year wait. For many of those customers, lead time was the prime factor in the decision to purchase a Japanese machine.

FIGURE 4 Machine Tool Industry—Cutting and Forming: Indices of Unfilled Orders and Imports (1967=100).

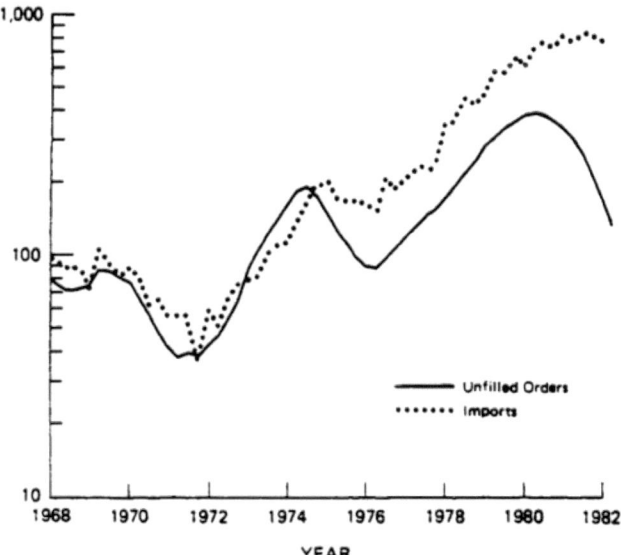

Source: NMTBA, Economic Handbook of the Machine Tool Industry 1982–83, p. 130.

Reliability. As in other areas such as electronics and automobiles, Japanese machine tools have gained a reputation for reliability. Respondents who were interviewed for this report stressed the superior reliability of Japanese machine tools over the American counterparts, and the meticulous attention to after-sales service.

A complete comparison of reliability characteristics between Japanese and U.S.-produced machine tools would require access to proprietary data. The Committee notes, however, that business realities normally force greater attention to reliability by the exporter than by the domestic manufacturer, in order to reduce the expense of maintaining a large, after-sales servicing force in a foreign country. In particular, the Japanese attention to quality is substantially at odds with the pressures on American business to maximize production, sometimes at the expense of quality—what one Committee member described as a "get-it-out-the-factory-door, we'll-fix-it-in-the-field" attitude.

This attitude, to the extent it describes machine tool industry management, has clearly hurt the industry. Product reliability has become one of the major selling points of Japanese machine tool products, according to prime defense contractor respondents who had bought Japanese tools in recent years.

Targeting. Japanese exporters have evidently concentrated on certain segments of the machine tool market, both product- and customer-defined. With regard to product, the Japanese have been most successful in selling numerically controlled machining centers and lathes to U.S. customers. Line 7 of Table 8, for example, illustrates the dramatic growth in the percent of the domestic lathe market which has been captured by imports.

This selectivity is deliberate. As the Japanese Study Mission report points out:

> If [Japanese machine tool manufacturers] find that the potential for market share does not exist, they will skip a product or model. Unlike many U.S. manufacturers, they will not manufacture a product just to round out the product line—they are very selective in machine sizing....[32]

The Japanese Machine Tool Builders' Association reported that 64 percent of its members' total NC machine tool shipments in 1980 went to small companies. U.S.

builders in contrast have tended to rely on larger, easier-to-serve customers such as manufacturers of automobiles, aircraft, farm equipment, and off-road vehicles.

Prices. Although price has sometimes been less important than delivery time and reliability, it is nevertheless a major factor in markets and a critical problem in view of the present underutilization of capacity in the U.S. industry. Japanese companies have been able to sell certain machine tools in the United States for 10 to 40 percent below U.S. producers' prices. As Table 9 shows, the Japanese price advantage is largely a "cost advantage" that plays a substantial role in Japan's competitiveness.

Table 9 provides a rough breakdown of the costs for building a conventional computer numerical control (CNC) lathe in the United States and Japan. While the table is intended only as an indication of general trends, the magnitude of the cost advantages enjoyed by the Japanese manufacturer is impressive. The data show a Japanese advantage at every step, despite estimates of a higher percentage for indirect labor. The resulting 21 percent price differential is typical of the experience of machine tool purchasers who were interviewed for this report.

The Committee found the following elements to be the primary contributors to this price differential:

- Purchased materials. In the table, the Japanese are shown to have a 30 percent cost advantage. That statistic, however, could reflect differences in the mix of "make vs. buy" decisions between Japanese and U.S. machine tool firms as much as it might reflect actual cost advantages. The Committee was unable to determine whether the Japanese machine tool industry may be more inclined to purchase a relatively small amount of materials (which might explain the lower figure for purchased material) and make a higher proportion of components in-house (which might explain the higher labor-hour figure).
- Dollar/yen exchange rate. The dollar-yen rate is a two-edged issue. Although the dollar is currently rather strong against the yen, giving Japanese manufacturers an across-the-board price advantage in U.S. markets, this strength is also responsible for attracting investment funds to this country in a way that has helped fuel the current economic recovery substantially. A

TABLE 9 Comparative Costs of CNC Lathe Tool: U.S. vs. Japan, 1981

	United States		Japan	
	Amount	Percent	Amount	Percent
Manufacturers' Selling Price	$120,000		$92,240	
Gross margin a/	48,000		36,900	
Manufacturing cost	72,000	100	55,340	100
Purchased material b/	32,400	45	22,680	41
Labor and burden	39,600	55	32,660	59
Direct labor c/				
Dollars	9,900	14	8,165	15
(Hours)	1,081		1,384	
Indirect and burden d/	29,700	41	24,495	44

a/ Gross margin of 40 percent is assumed for both U.S. and Japanese producers.

b/ For the U.S., purchased materials are 45 percent of manufacturing cost; for Japan, the cost is 30 percent less than the U.S. material cost.

c/ For the U.S., labor cost is estimated on the basis of a 1 to 3 ratio between direct labor and indirect labor and burden. Unit hours are derived by dividing direct labor cost by 1981 average hourly earnings of production workers in metal-cutting machine industry ($9.16). (U.S. Bureau of Labor Statistics) For Japan, direct labor hours per unit are derived by increasing U.S. levels by 28 percent, in accordance with 1980 estimates by the Japan Productivity Center of comparative levels in the industrial machinery indsutry. The 1981 hourly average for Japan is $5.90.

d/ Indirect and burden are derived as residuals. The higher proportion for Japan (despite lower fringe benefits) reflects the higher ratio of non-production workers to all employees in Japan's metalworking machinery industry (40 percent) compared with the U.S. industry ratio (30 percent), according to BLS data.

Source: U.S. Bureau of Labor Statistics, Japan Productivity Center, and Committee calculations.

- premature stifling of these flows could severely damage the recovery, and the spillover effects of this would harm U.S. machine tool builders.
- Productivity. Japanese productivity growth has been substantially above that of the United States. During 1973–1981, Japanese manufacturing output per man-hour grew approximately 8 percent annually, compared to an average annual decline in output per man-hour of 0.7 percent in the U.S. machine tool industry. Because Japan started from a lower output-per-man-hour base, Japanese overall productivity still lags that of the United States. Productivity growth is an important component in the competitiveness of the industry, however, as it has a direct link with the industry's levels of capital investment. As a general rule, those industrial sectors that enjoy more rapid productivity growth and are associated with larger amounts of capital investment also enjoy greater price stability than the slower-moving sectors.
- Superior machine tool manufacturing facilities in many cases (i.e., more modern, more highly automated, etc.).

<u>Commercial and Government Policies Regarding Industrial Development</u>. The Japanese approach to industrial development has been an important aspect of Japan's postwar economic success, and has given rise to the expression "Japan, Inc." Some of the key elements of this system are:

- close industry-government cooperation in planning industrial development
- less restrictive application of antitrust laws, with the effect of allowing vertical integration of larger companies, and horizontal coordination among actual and potential competitors for R&D, product specialization, and other purposes
- financial practices that allow higher debt-equity ratios than would be prudent in the United States and, thus, greater access to credit
- government-encouraged financial support
- close cooperation by labor with its associated industrial company

The effects of these policies and practices are difficult to assess. Observers who are familiar with both the Japanese and U.S. business environments assert, however, that the Japanese "system" in the aggregate provides advantages the United States simply may not be able to match under this country's present customs and labor-management-government relations. By contrast with the Japanese, U.S. companies operate under a less cohesive system, characterized by more restrictive antitrust laws, frequently adverse industry-government and industry-labor relations, uncertain national purpose, and less advantageous financial conditions.[33]

A final observation regarding the effects of governmental policies concerns the management of the macro-economy itself. The economic characteristics of an industry such as machine tools are not completely independent from the characteristics of the overall economy. For a number of reasons, the slow growth of the American economy in the last decade expressed itself in a sluggish demand for machine tools—as for capital goods generally. The weak demand for machine tools has been a significant factor in the slow productivity growth in the machine tool industry itself. Conversely, in Japan, rapid growth in aggregate output has been accompanied by higher rates of investment and more rapid productivity growth in machine tool production. Thus, to some extent, the performance of each country's machine tool sector has been consistent with the differential growth rates of each economy.

Because of the key role of foreign competition in determining the long-term survival of the U.S. machine tool industry, the Committee has examined the machine tool policies in three countries: Japan, France, and West Germany. These are presented in Appendix B.

NEW ENTRANTS AND NEW COMPETITIVE STRATEGIES

The technological advances and the global nature of machine tool competition, described above, have caused a number of changes in (1) the types of competitors in the broadly defined machine tool market and (2) the competitive strategies that will be required by those selling in this market. Together these changes raise important issues affecting the longer term health of the American machine tool industry. This section looks at those issues in terms of the DOD interest in maintaining a healthy, across-the-board domestic machine tool productive capability. It is based on the Committee's finding that new competitive conditions will require new qualities and skills of U.S.-based builders.

The major changes taking place in the competitors and competitive strategies, including the issues that these changes raise, are threefold:

(1) The development of computer-integrated manufacturing has attracted large, U.S.-based, multinational firms to the market for products used in automated manufacturing that are ancillary to machine tools. These companies have not entered the business of manufacturing machine tools themselves, and it is unlikely that they will do so in the near future. If they undertake to supply an FMS or automated factory customer with machine tools, they will probably purchase the tools from a machine tool manufacturer. However, as the markets for FMS and other factory automation systems develop, these new entrants will be formidable competitors with machine tool producers for the "ancillary" products needed in factory automation—which in many systems will be of greater value than the machine tools themselves. Indeed, some of the new entrants have gained experience in automating their own facilities, and are well positioned to compete successfully in the new technology of factory automation. A few of them, alone, have greater financial resources than the entire traditional U.S. machine tool industry combined. They also have had extensive experience in international trade, including international joint ventures. Moreover, by their machine tool purchasing decisions, these companies may determine whether a substantial portion of the machine tools consumed in the United States is produced here or overseas.

- Will the entry of these larger firms change the "rules of the game," making it even more difficult for smaller, traditional machine tool firms to compete?
- Will they necessarily turn to U.S. machine tool makers to supply the basic metal-forming and metal-cutting equipment for their technology?

(2) Another set of entrants comprises small entrepreneurial firms dedicated to relatively narrow, high-technology product lines related to machine tools. Many such firms have already developed reputations for quality in software, customer support, customer training, and applications engineering (i.e., the combination of services needed to support computer-integrated and flexible manufacturing systems), as well as robotics. Experiences in Japan and Germany suggest that tech

nologically innovative, small firms can compete quite effectively with larger firms if given reasonable access to R&D funds and credit.

- Does the existence of such firms offer significant potential for the U.S. machine tool industry to remain a international leader in new manufacturing technology?

   (3) The structure of the industry is changing, with the solidification and further development of "strategic groups"[34] based on new categories of machine tool production (e.g., robot systems, integrated manufacturing systems). Even traditional strategic groups (e.g., stand-alone machine tool builders) are being required to adopt new strategies, such as locating facilities abroad in order to survive.

- Will the new and the traditional strategic groups each contain, and will they retain, adequate domestic productive capacity to ensure a healthy, competitive industry capable of serving DOD's needs?
- To the extent that domestic machine tool makers themselves branch out into overseas production for consumption in the U.S. market, will this help or impair U.S. defense readiness?

In concluding that new competitive conditions will require new qualities and skills from U.S. machine tool builders, the Committee observes that developments in world machine tool competition are being driven by two major forces: (1) technology advances in factory automation and materials processing, and (2) an increased need for customer support, primarily in the form of engineering services required from the supplier to match services supplied by foreign suppliers and the increasing sophistication of machine tool products. These two criteria can be used to "map" the various strategic groups in the machine tool industry today. Figure 5 contains such a map of the industry today, with major machine tool product categories placed according to their relative sophistication of technology and degree of customer support. The two axes—technology, and sophistication of customer support—help define both the strategies of the groups and the criteria for survival in each one.

In the lower left corner of the map are stand-alone machine tool makers. In contrast to the products of more technologically sophisticated firms, the products of this group require relatively less customer support and information systems technology. Firms marketing these products, therefore, will compete mostly on the basis of price, delivery time, and reliability. Because this is one group where Japanese manufacturers have tended to compete heavily, competitive conditions will probably require U.S. firms in this group to become competitive

FIGURE 5 Strategic Groups

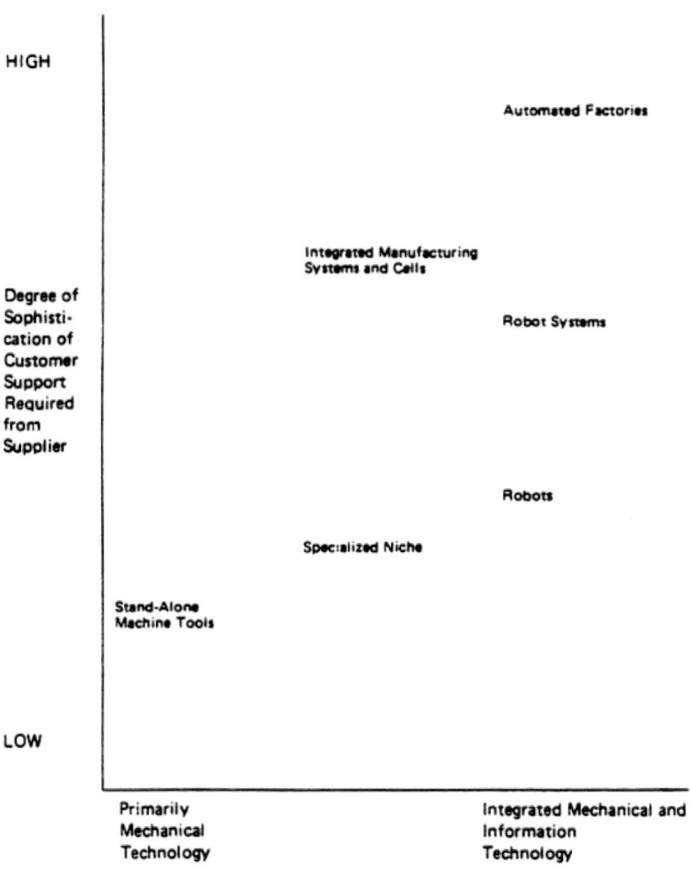

worldwide like the Japanese, to reach small as well as large customers, and to develop economies of scale in production.

The manufacturers of integrated manufacturing systems, the group in the middle of the map, face somewhat more difficult technological as well as marketing problems. Because this is an area where the range of possible applications has not yet been fully developed, participants in this group must divert substantial (relative to the stand-alone manufacturers) resources to R&D. Field interviews for this report revealed that many such producers must overcome user skepticism over the prospect of purchasing a highly automated machine, and over the confusing array of options, the fear of down-time, the programming and maintenance costs, the integration of new machines with existing production control systems, and the compatibility of the new machines with future adaptations. This means that marketing and after-sales servicing will need to accommodate the first-time user, all of which requires greater efforts at customer support.

The number of potential manufacturers thins out quickly once one leaves the lower left corner of the map. Relatively few conventional machine tool companies have the range of necessary skills to compete successfully in the middle group, which currently comprises primarily the large machine tool manufacturers who account for a sizeable portion of U.S. machine tool production capacity. It may well be that this group is, as a practical matter, open only to larger firms that have the resources to offer more comprehensive after-sales service and to gain better access to capital markets.

Greater resources are required to sell entire, automated plants. To the skills required by the producers of integrated manufacturing systems, one must add major project management and strong information technology capability. Projects of this kind require the ability to weather long selling cycles, and also require strong customer support and planning abilities. Larger firms are likely to have an advantage here as well, but only when they are able to develop truly coordinated systems (consulting, production, marketing, etc.) within their companies. As yet, no U.S. firm has built and sold a complete, fully integrated batch manufacturing plant. Such complex facilities, however, are being sold in other parts of the world, often by consortia of companies and in some cases by large, government-subsidized companies such as Renault.

A few niche positions remain, such as in precision and very high speed machine tools. In these areas, a small, high quality maker can focus attention on special market needs. Such a firm may still better serve a specialized market than a large firm addressing broad segments of the industry. On the other hand, many of the former specialty slots will disappear, undercut by the ability of new competitors to provide high performance, multi-purpose systems at modest cost.

Table 10 compares the skills described above with the characteristics of the traditional machine tool industry which was analyzed earlier in this chapter. While the comparison necessarily deals in generalities, the directions they lead the observer are clear: each new strategic group requires skills substantially different from, and more complex than, the ones which were adequate for competing in the traditional machine tool business.

The ability of the broadly different domestic machine tool industry to respond to defense needs rests, in part, on the makeup of the industry at a particular time. One might reason, for example, that a rapid growth in the number of domestic robot manufacturers signals a healthy response of the U.S. industry to developments in the market for robotics equipment and accessories. Failure of the industry to move into new product areas, on the other hand, could reasonably be interpreted as an indication that U.S. firms were having some difficulty adapting to new market realities.

Table 11 sets forth rough estimates, based on NMTBA. member responses and information obtained from the Robotics Institute of America, of the number of U.S. firms in each strategic group. Although the figures set forth in the table must be interpreted as "soft" (they are based upon voluntary membership responses and are not checked for consistency), they do give some idea of the trend of machine tool manufacturing: The number of manufacturers that claim to be venturing into sophisticated machine tool technology is growing, consistently with the growth of markets for new types of machine tools.

Although the table can give some cause for hope that U.S. machine tool firms can perform adequately across the range of necessary technology and customer support, it also raises some cause for concern. Many firms remain at the lower left end of the map. These constitute the bulk (by number of companies but not by volume) of the traditional machine tool industry which was analyzed at the beginning of this chapter. Such firms represent a potential resource, but only if they are capable of adapting to new competitive conditions.

TABLE 10 Skills Required in the Emerging Machine Tool Industry

	Profitability/ Finance	Manufacturing	Employment Patterns
Traditional U.S. Machine Tool Company (includes manual machines and single-station special-purpose machines)	Maintain moderately healthy financial ratios while remaining relatively small as a firm	Manage sales fluctuations; "Buffering"	Manage employment fluctuations
New, Stand-Alone Machine Tools	Financial management must support more investment R&D	Achieve greater competitiveness by stressing volume output	Concentrate on production economies through labor savings
Integrated Manufacturing Systems	Help users finance systems; maintain investment during downturns.	Integrate diverse parts; use stand-alone tools efficiently	More complex compensation system
Automated Whole Plants or Custom Lines	Long selling cycle requires substantial financial resources; help customers with innovative finance, esp. exports	Understand/make/ buy/integrate diverse parts	Functional career planning workplace stability required

As the next chapter points out, DOD has identified some bottlenecks in the surge production of certain weapons systems. The response of the domestic machine tool industry will influence whether these bottlenecks will worsen or can be resolved in the long run. The next section of this chapter examines the response of machine tool builders to new competitive conditions, and examines some of the issues raised by this response.

RESPONSE OF MACHINE TOOL BUILDERS TO THESE CHANGES

The preceding sections of this chapter have described changes taking place in machine tool manufacture, markets, and industry structure, and have defined the skills and qualities that will be required for U.S. suppliers if they aim to meet new competitive conditions. How the U.S. machine tool industry responds is, of course, critically important to its survival and to the country's national security interests.

Capital Investment	Research and Development	Productivity Management	Marketing
Low capital investment levels possible; increase production through employment increases.	Low R&D levels; application engineering forms substantial portion of R&D	Low productivity levels; buffering practices place low premium on latest production technology in own operations	Technology is customer-driven; relatively low levels of customer support required.
Maintain higher levels of capital investment	NC controls, new products require more R&D	Substantially higher levels of productivity growth required	Programs for large and small customers
Investment levels must support high-tech standards	System integration, sensors, etc. * more R&D	Strong international competition places premium on high productivity	Sophisticated selling; uUser education
Investment levels must support high-tech standards and massive projects	Multiple R&D efforts, through own labs, universities, government	Strong international competition places premium on high productivity	Government contracting capability; vVery sophisticated customer support

To gain some appreciation of this response, the Committee compiled a list of recent economic and technological trends shaping the machine tool industry. Using a questionnaire, it then asked NMTBA members to comment on the extent and ways these trends were affecting their individual firms and the industry as a whole. In all, 43 NMTBA members responded to the 100 questionnaires that were issued. The Committee also visited several machine tool builders, to interview their executives at greater length about trends affecting their companies. Eleven such site visits were made.

TABLE 11 Estimates of the Number of U.S. Machine Tool Firms in Specified Strategic Groups

Strategic Group	1974	1977	1982	1983
Stand-alone machine tools	223	253	N/A	285
Specialized niche machine tools	56	71	N/A	75
Robots	4	N/A[a]	N/A	39
Robot systems	N/A	N/A	N/A	12
Integrated manufacturing systems and cells	N/A	N/A	29[b]	37[b]
Automated factories	0	N/A	[b]	[b] [c]

Sources: NMTBA Directories; Robotics Institute of America.

[a] According to the Robotics Institute of America, there were 4 U.S. robot manufacturers in 1974, and less than 10 manufacturers through 1977.

[b] Approximately 65 NMTBA members reported for the 1983 Directory that they manufactured computer controlled machinery or computer controls, up from 45 in 1982.

[c] At least one U.S. machine tool firm is constructing an automated factory. At least 2 others manufacture the range of products required for such construction.

The Committee found that a variety of actions characterize the industry response to new competitive conditions. These are described below (see "Coping With Change—Specific Steps").

Economic Pressures

By and large, the 43 machine tool builders responding to the survey considered the economic trends to carry more serious consequences than the technological ones. Concerns about present economic health—and in many cases survival—appeared to overshadow concerns about the role of technological leadership in remaining competitive.

In view of the evident vigor and resolution with which Japanese machine tool builders are applying the latest technology, this attitude—while understandable—was worrisome to Committee members. It suggested that extraordinary efforts might be required among American machine tool builders in order to maintain their reputation for technological excellence.

Increasing competition, including price competition, from foreign manufacturers in both foreign and domestic markets was ranked by all 43 machine tool builders as being of highest importance to the industry. Two respondents, however, said that foreign competition had little impact on their own firms. This seeming anomaly was explained by the Japanese "targeting" of such products as machining centers, to the exclusion of others; producers of some specialized machines have found successful niches.

The high cost of capital over a prolonged period has been a double-edged problem for the industry. The number of machine tool orders has dropped, as potential customers are unable to finance major purchases; and the borrowing power of the machine tool firms themselves has dropped recently with decreased sales and profits.

As machine tool orders pick up during the current recovery, there is considerable apprehension among U.S. builders that their position will be further weakened by the inventory of Japanese tools presently stored in U.S. warehouses. According to the petition filed by the NMTBA, Japanese inventories of NC lathes and machining centers in this country stand at the equivalent of 1–1/2 years production for NC lathes and 9 months production for machining centers.[35] This description about Japanese machine tool inventories in the U.S. has been disputed in a response to the petition.[36]

Coping With Change—Specific Steps

The site visits and questionnaire responses revealed that U.S. builders are using a variety of approaches and strategies to become more competitive. Some of these actions, however, raise questions about their longer-term effects on the national security. In two categories, mergers and joint ventures, the Committee considers the issues serious enough to warrant further investigation.

The following description of competitive steps being taken by respondent firms—while not exhaustive—gives some indication of the "shakeout" taking place in the U.S. machine tool industry today. As in all shakeouts, there will be survivors and those that do not survive. Those survivors that continue to manufacture and sell machine tools profitably will necessarily be more forwardlooking firms, committed to seeking global sources and markets.

<u>Conventional Cost-cutting</u>. These steps include layoffs and furloughs, dividend cancellations, union contract renegotiation, and liquidation of facilities in high-cost locations in order to move manufacturing operations to lower-cost areas in the U.S. or abroad. The NMTBA, for example, has identified approximately 40 U.S. machine tool firms with manufacturing facilities outside the United States. Most of these facilities have been used to penetrate foreign markets, especially in Europe. However, at least one industry analyst cites as a "trend" the move toward U.S. firms' involvement in overseas production of machine tools for U.S. consumption.[37] This subject is covered in further detail under "Joint Ventures," below.

<u>Reorientation of Business Strategy</u>. At least one large machine tool firm has pledged to "out-Japanese the Japanese." It has instituted Japanese methods in inventory management, quality control, marketing strategies, and customer service, as well as an emphasis on quality and the adoption of FMS technology for its own production.

Some traditional U.S. machine tool companies are diversifying into the production of plastics forming machines, robots, microcomputer components, software turnkey services, and materials handling systems.

Some new firms have attempted to identify markets (e.g., certain types of controllers) where both Japanese and U.S. competition seems weak. In one successful case of "niche-playing" the firm involved has been able to maintain relatively even growth, in spite of sales fluctuations in the machine tool markets as a whole. (For more information on niche-playing, see "New Entrants and New Competitive Strategies," above.)

Efforts to Gain New Technological Expertise. These include budgeting additional R&D sums for discovering new technology (see, however, the discussion of R&D spending in this chapter), pursuing some contracts on a "breakeven" basis in order to gain experience in useful technology, and making minority investments in firms that have expertise in relevant technologies.

Mergers and Acquisitions. These have been common for some time as a strategy to remain competitive. Recent examples include the Cross Company's merger with Kearney and Trecker, the acquisition of Unimation (a maker of robots but not of traditional machine tools) by Westinghouse (a maker of industrial controls and a seller of factory automation services but not a manufacturer of traditional machine tools), the growth of Newcor and Lamb Technicon through acquisitions, the acquisition of Snyder by Giddings and Lewis, and the subsequent acquisition of Giddings and Lewis by AMCA International. In the case of the Cross/Kearney and Trecker merger, the U.S. Department of Justice diluted the possible competitive benefits by requiring that the merged company divest itself of certain product lines.

Mergers and acquisitions hold out the possibility for economies of scale and the ability to attract sufficient funding for necessary capital improvements. If managed properly, a machine tool firm involved in a merger or acquisition could enjoy the benefits of a stronger capital structure, better access to R&D funds, and possibly an international sales and administrative structure. All of these are essential for competing successfully in a modern, global machine tool market. The Committee has two concerns regarding such developments, however, regarding the ability of the merged or acquired machine tool firm to compete. (1) The joining of a domestic machine tool firm with a larger non-machine tool entity could result in severe cost cutting, the use of the acquired firm's liquidity to finance other initiatives within the parent corporation, and the imposition of a large corporate bureaucracy; all these are common effects of mergers and acquisitions today. (2) The joining of a domestic machine tool firm with a foreign firm that intended to use its U.S. base chiefly as a sales outlet could strengthen the domestic firm's short-term financial structure at the expense of an ability to design and manufacture its own products.

If these effects became characteristic of mergers and acquisitions within the machine tool industry generally, the merger/acquisition movement—instead of enabling individual machine tool firms to maintain their competitiveness—would bring few improvements to the domestic industry. The Committee believes that this possibly harmful aspect of mergers and acquisitions on U.S. machine tool manufacturing capability is an important issue worthy of additional study.

Joint Ventures. Many firms are finding that the most efficient route to gaining access to additional skills and product lines is to pursue joint ventures, especially with foreign partners. Joint ventures are common among companies trying to reposition themselves strategically. Examples include Bendix-Murata, Acme Cleveland-Mitsubishi, Westinghouse-Mitsutoki, General Motors-Fanuc, and Rockwell International-Ikegai Iron Works. Clearly, many major players are involved.

Most of these joint ventures have offered the potential for low-cost, reliable overseas manufacturing for the U.S. partner, and an enhanced marketing network in this country for the foreign one. They represent the trend toward global sources taking place in the industry. They raise some questions, however, as to the effect that such actions could have on the long-run competitiveness of machine tool manufacturing facilities located in the United States. When Bendix acquired Warner and Swasey, for example, one of its first actions was to transfer nearly all of its machine tool production to the Murata joint venture in Japan. Subsequently, Acme-Cleveland has announced that its state-of-the-art NC chucker, jointly developed with Mitsubishi Heavy Industries, Ltd., will be produced in Japan,[38] and Cross and Trecker has said that it is not committed to the production of any percentage of its machine tools domestically.[39]

Concern was expressed by the Committee that if the practice of overseas procurement or production by U.S. companies of machine tools for sale in the United States were to become widespread, there would be the long-term danger that U.S companies would end up more as distribution channels for foreign-built machine tools than as manufacturers in this country.

Requests for Federal Assistance

Two recent petitions by machine tool builders to the federal government for relief from the competition of foreign machine tools represent another kind of response to the economic trends that have been described. It is not within the Committee's charter to pass judgment on these petitions. However, because they are relevant, we note them below.

The first petition was submitted on May 3, 1982, by Houdaille Industries, Inc., to the Office of the U.S. Trade Representative, asking for the President to exercise his authority[40] to deny the benefits of investment tax credits when producers have an unfair price advantage as the result of a cartel. Attorneys for Houdaille Industries conducted extensive research to document practices in Japan that could be construed as contributing to a machine tool cartel. Some of their evidence is incorporated in the Japan section of Appendix B. The petition was denied in April 1983.

A second petition is pending as of this writing. The National Machine Tool Builders' Association has submitted a petition to the U.S. Department of Commerce under the National Security Clause, Section 232 of the Trade Expansion Act of 1962 (19 U.S.C. Section 1862). This petition requests a five-year period during which imports of both metal-cutting and metal-forming tools would be limited to 17.5 percent of the value of total domestic consumption. The argument for this action is that "the national security of the United States is being impaired by current levels of imports of machine tools because such imports threaten to debilitate the domestic machine tool industry, which is critical to the United States' defense and deterrence posture."[41]

CONCLUSIONS

This chapter has described a changing machine tool market which, in the course of five to seven years, has become significantly more competitive and complex.

- Advances in microelectronics, robotics, systems engineering, computer science, and substitute materials have altered the character of manufacturing and changed the nature of the machine tool industry, making machine tool construction (as defined in this report) one of the

world's "high tech" industries. Further advances in the commercialization and military application of synthetic materials that substitute for metal will also affect manufacturing technology and, ultimately, the size of the market for conventional machine tools.
- International competition, especially from the Japanese, has brought intense pressures on U.S. firms to meet new standards of innovation, reliability, price, and customer service.
- New entrants to the market for automated manufacturing have brought new (to the traditional machine tool industry) specialties such as computers and software for design and integration; electronic controls; machines for assembling, testing, plating, and heat-treating components; robots; and sophisticated engineering services. These firms have combined resources that could expand the financial power of the manufacturing process industry by several times the present size of the machine tool industry as traditionally defined. In addition, conglomerates such as Allied-Bendix, Litton, Textron, White Consolidated, and AMCA International have substantial machine tool subsidiaries. The actions they take to rationalize their machine tool operations may accelerate the already rapid change in the industry, providing they invest in strengthening their machine tool elements. The financial power of these new firms, and their "high-tech" orientation, may require smaller firms to merge in order to become large enough to make the investments now required to remain competitive in a technologically advanced industry.
- New strategic groups in the industry have relegated many traditional machine tool producers to the "lower left" spectrum of an industry map that ranks strategic groups according to the degree of technological sophistication and customer support required. The traditional machine tool firms produce in an environment in which their products are more like commodities than products of greater technological sophistication requiring extensive computer and other engineering services. In this traditional market sector, which is now actually part of a larger machine tool market, this strategic group will have to adjust its capabilities to meet intensified competition on the basis of price, delivery time, and reliability: factors where such U.S. firms have shown comparative weakness in recent years.
- The globalization of machine tool manufacture and markets has forced U.S. machine tool builders themselves

to take a global view of sources and markets, including the location of manufacturing facilities overseas.

These new realities require skills and characteristics substantially at odds with the description of the traditional machine tool industry on pages 8 to 18 of this report.

Thus, the signs of a far-reaching "shakeout" in the machine tool industry are unmistakable. While some domestic machine tool builders will be unable to respond to increased competitive pressures from abroad and from alternative technologies, there are a number of forwardlooking firms—among them traditional machine tool builders as well as new entrants to the market for products and services ancillary to the use of machine tools in automated manufacturing applications—that have recognized and reacted to the trends that are forcing changes. Those domestic firms that have had the foresight to move toward automated systems development, processing of non-metals, advanced machining and forming techniques, and a global view of markets (or, in some cases, successful niche-playing) will survive despite a continuing, substantial challenge from foreign producers. These firms will continue to be able to respond to the needs of the Department of Defense.

For some domestic machine tool builders, however, the economic trends—high cyclical demand followed by the especially sharp downturn of the recent recession—have had two consequences that may well be fatal. First, the effects of economic cycles have distracted some machine tool builders from the fundamental technological changes that are proving to have a lasting impact on the types of products and services demanded, and on their own manufacturing methods. Failure to respond to those changes has left a number of firms with product lines which, because they incorporate less sophisticated technology or because they employ traditional manufacturing methods, must now compete fiercely on the basis of price, delivery time, and reliability, which they have proven ill-prepared to do in the past. Second, new competitors from abroad have made inroads into the machine tool market that are unprecedented despite the history of cyclical machine tool demand.

While the evidence of a structurally more dynamic industry is welcome, two trends raise questions about the benefits, from a national security standpoint, of changes that are taking place:

(1) Sound business decision making today may dictate that a corporation shift its machine tool production to a foreign venture partner, seek foreign machinery to complement its own peripheral devices such as controllers, or relocate its own manufacturing facilities overseas. The danger exists that as business comes closer to realizing true economies of production on a worldwide scale, the United States could lose some productive capacity which is valuable to the national security.

(2) Among the responses of traditional machine tool builders to increased competition has been a request for limited, temporary protection from imports. The danger exists that efforts to provide immediate help for domestic machine tool builders will, without vigorous and successful efforts by the industry to improve its own productivity and technological position, actually weaken that industry's ability to provide the leading-edge technology and to compete successfully on a global basis.

To deal with such issues requires an understanding of how the Department of Defense, prime defense contractors, and the machine tool industry interact. The next chapter examines these subjects.

NOTES

1. National Machine Tool Builders' Association (NMTBA), Economic Handbook, 1982–1983, p. 1.

2. Petition Under the National Security Clause, Section 232 of The Trade Expansion Act of 1962, for Adjustment of Imports of Machine Tools, Submitted by National Machine Tool Builders' Association ("Petition"), p. 20.

3. U.S. Bureau of Labor Statistics, Employment and Earnings, July 1983, pp. 78–81.

4. Anderson Ashburn, ed., "World Machine-Tool Output Falls 20%," American Machinist, Feb. 1983, p. 77.

5. NMTBA, Economic Handbook, p. 164.

6. NMTBA, Economic Handbook, p. 167.

7. National Academy of Engineering, "Competitive Status of the U.S. Machine Tool Industry," 1983, p. 18.

8. U.S. Department of Commerce, Census of Manufactures, 1977.

9. American Iron and Steel Institute, Annual Statistical Reports.

10. "Petition," page 35.

11. See, for example, NMTBA, Economic Handbook, p. 249.

12. It was reported on May 27, 1983, that Cincinnati Milacron and Acme-Cleveland had reported three consecutive quarterly losses. Gleason Works had reported five and Brown & Sharpe had reported six. Value Line (Machine Tools), May 27, 1983, p. 1344. In 1982 annual reports it was stated that Lodge & Shipley lost money during 1982, as did the machine tool segments of Textron and White Consolidated Industries.

13. Eli S.Lustgarten, Vice President, Paine Webber Mitchell Hutchins, quoted in American Metal Market, June 13, 1983, p. 2A.

14. Otto Hintz, et al., "Machine Tool Industry Study Final Report," U.S. Army Industrial Basic Engineering Activity, Rock Island, Illinois, November 1968, pp. 18–19.

15. National Machine Tool Builders' Association, "Meeting the Japanese Challenge," 1981, p. 7.

16. National Academy of Engineering, op. cit., p. 49; Otto Hintz et al., op. cit., p. 36.

17. "Petition," p. 132.

18. NMTBA, Economic Handbook, pp. 162–163.

19. See, e.g., U.S. Congress, Joint Economic Committee, Special Study on Economic Change, Volume 3 Research and Innovation: Developing a Dynamic Nation (1980), passim.

20. See, e.g., NMBTA, "Meeting the Japanese Challenge," p. 13.

21. Cincinnati Milacron, Annual Report, 1982.

22. Testimony of W.Paul Cooper, Chairman, Acme-Cleveland Corporation, before the International Trade Commission on June 28, 1983, p. 3.

23. NMTBA, "Meeting the Japanese Challenge," p. 14.

24. See notes 9, 10, 11 of Chapter 3 of this report.

25. NMTBA, "Meeting the Japanese Challenge," p. 14.

26. Joint Logistics Command, Heavy Press Study, November 23, 1982.

Douglas Aircraft reports that in the next 15 years there will be no major application for composites in large airframe structures due to the many existing technological problems. This position was also supported by Rockwell-International. Although they are using composites (up to 700 pounds of parts) per B-1 aircraft, they are not among the "large bones" of the structure. Boeing feels that as the technology develops, an increasing percentage of aircraft will be made up of composites.... Many of the current generation of composites (i.e., those containing graphite) are unacceptable in Naval surface ship combat environment since debris from composite damage could affect EMI/EMC.

27. Cincinnati Milacron, Annual Report, 1981.

28. NMTBA, Directory 1983, pp. 33–94.

29. Ibid., p. 21.

30. Testimony of Eli Lustgarten, Vice President, Paine Webber, Mitchell Hutchins, before Economic Stabilization Subcommittee of House Committee on Banking, Finance and Urban Affairs, July 26, 1983, p. 1.

31. NMTBA, Economic Handbook, p. 167.

32. NMTBA, "Meeting the Japanese Challenge," p. 27.

33. For a good historical review of the Japanese "system," see U.S. Congress Office of Technology Assessment, U.S. Industrial Competitiveness: a Comparison of Steel, Electronics, and Automobiles (Washington, 1981) pp. 190–193.

34. Strategic groups are groups of manufacturers which, for reasons relating to the similarity of product or market, follow similar business strategies. The concept of strategic groups comes from Michael Porter; see, e.g., Porter, Competitive Strategy (New York, The Free Press, 1980).

35. "Petition," pp. 153–154.

36. Japan Machine Tool Builders' Association response to NMTBA Petition, "Investigation of Imports of Metal-Cutting and Metal-Forming Machine Tools under Section 232 of the Trade Expansion Act of 1962," pp. 136–142.

37. Lustgarten, op. cit., p. 19.

38. Cooper, op. cit., p. 10.

39. Testimony of Richard T.Lindgren, President and Chief Executive Officer, Cross & Trecker Corporation, before the International Trade Commission, June 28, 1983, at p. 8.

40. Section 103 of the Revenue Act of 1971, 26 U.S.C. Section 48 (a) (7) (D).

41. "Petition," p. 6.

3

THE DEPARTMENT OF DEFENSE, PRIME CONTRACTORS, AND THE MACHINE TOOL INDUSTRY: RELATIONSHIPS THAT AFFECT INDUSTRY STRUCTURE

This chapter examines relationships among the Department of Defense, the domestic machine tool industry, and those prime defense contractors that are the major users of machine tools. These relationships include not only such direct mechanisms as contracting procedures, but also the attitudes and perceptions that affect the ability of one party to work with another. The Committee has found that such attitudes and perceptions ultimately affect industry structure.

The circumstances in which conventional machine tool manufacturers now find themselves, described in the previous chapter, are obviously only partly attributable to characteristics of the defense market. Therefore, the machine tool firms cannot be changed in any major way by DOD actions alone. Indeed, the major forces for changing the industry are not defense-oriented. Nevertheless, the Committee believes that the Defense Department's direct and indirect influence on the industry can be substantial.

Although DOD direct purchases of machine tools are small compared with total domestic machine tool production, DOD's influence on industry behavior manifests itself indirectly, through the requirements placed on prime contractors. In fact, the prime contractor role in the DOD-contractor-supplier triangle has sometimes been likened to a buffer between the small supplier on the one hand and the government (with its burdensome contracting procedures) on the other. As discussed below, the defense sector remains a significant market for the products and services of machine tool builders.

The following pages analyze the size of the DOD and defense prime contractor market for machine tools and focus on two distinct Defense Department roles in that market: DOD procurement, and DOD support of technology development and application. The chapter then considers the prime contractors' view of the machine tool industry, followed by a review of legislation affecting domestic machine tool purchases.

SIZE OF DOD AND CONTRACTOR MARKETS

The Department of Defense is by itself a rather small purchaser of machine tools, accounting for approximately 3.5 to 4 percent of domestic orders in 1978,[1] compared to the automotive industry's 28–30 percent and the civilian aerospace industry's 10–12 percent. An earlier (1972) estimate of the proportion of machine tool sales accounted for by defense contracts in total is 7.1 percent, indicating that purchases by private defense contractors were roughly equal to those made directly by DOD.

More recent estimates derived from an input-output analysis by the Commerce Department's Bureau of Industrial Economics (BIE) confirm this general level of DOD involvement.[2] The BIE concluded that in 1982 purchases by the Defense Department and its contractors together accounted for 6.2 percent of domestic metal-cutting machine tool production and 4.8 percent of metal-forming machine tool production. Assuming adoption of the Administration's 5-year defense plan and realization of the Council of Economic Advisers' projections for economic growth, the BIE estimates that the comparable figures in 1987 will be 7.5 percent and 6.3 percent, respectively.

A similar analysis conducted by Data Resources, Inc. (DRI), for the National Machine Tool Builders' Association[3] shows a much higher proportion of machine tool consumption when all indirect DOD supplier links (i.e., through prime contractor intermediaries) are considered. In Table 12, DOD "direct" purchases include tools for government arsenals, shipyards, and other defense installations. DOD "indirect" purchases include those by private parties on current account for delivery to defense agencies. Finally, "induced capital" purchases consist of those by defense contractors, subcontractors, and suppliers for use in the production of all military weapons and equipment. DRI concludes that "by conservative estimate, up to 20 percent of the aggregate domestic consumption of machine tools is related to defense needs even in peacetime."

TABLE 12: Domestic Consumption of Machine Tools

	Billions of 1972 Dollars						Annual Growth Rate
	1977	1978	1979	1980	1981	1982	1977 - 1982
Aggregate Consumption of Machine Tools	2.196	2.717	3.265	3.358	3.362	2.819	5.122
Aggregate Consumption of Metal-Cutting Tools	1.630	2.084	2.530	2.716	2.827	2.354	7.625
Aggregate Consumption of Metal-Forming Tools	0.566	0.633	0.735	0.643	0.534	0.465	-3.842
Aggregate Defense-Related Consumption of Machine Tools							
Direct	0.223	0.255	0.325	0.364	0.571	0.564	20.401
Indirect	0.080	0.083	0.088	0.093	0.265	0.280	28.472
Induced Capital	0.027	0.028	0.036	0.037	0.051	0.046	10.775
	0.116	0.144	0.202	0.234	0.254	0.239	15.592
Defense-Related Consumption of Metal-Cutting Tools	0.177	0.205	0.262	0.302	0.489	0.486	22.468
Direct	0.066	0.068	0.072	0.077	0.231	0.242	29.739
Indirect	0.020	0.021	0.027	0.029	0.042	0.038	14.011
Induced Capital	0.091	0.116	0.163	0.196	0.217	0.206	17.789
Defense-Related Consumption of Metal-Forming Tools	0.046	0.050	0.063	0.062	0.081	0.078	10.881
Direct	0.014	0.015	0.015	0.016	0.035	0.038	21.754
Indirect	0.008	0.008	0.009	0.008	0.008	0.008	-0.033
Induced Capital	0.025	0.028	0.038	0.038	0.038	0.033	5.651

Source: Data Resources, Inc., DEIMS and DIFS models (1983).

The DRI table also indicates that the defense share of the domestic market has grown as commercial sales have declined and remained depressed, even as the economy emerges from the recession. This increase may be attributed to both production increases entailed in the defense build-up and efforts to modernize DOD production facilities, including munitions arsenals and shipyards.

An important caveat is that none of the estimates takes into account the broader range of manufacturing equipment and systems, including related software, that should be considered along with the traditional categories of metal-cutting and metal-forming machine tools in assessing either DOD needs or the competitive status of the domestic industry. It is reasonable to conclude, however, that the defense sector remains an important market for these products and services and as such represents a far from negligible influence on the development of the domestic machine tool industry broadly defined.

DOD PROCUREMENT: INCENTIVES AND DISINCENTIVES

Department of Defense procurement begins with the preparation of a statement of requirements, usually two to three years in advance of funding and perhaps as long as four to five years before the equipment is installed and operating. The military services are required to search their own inventories before deciding to purchase new equipment. On the whole, these inventories contain older, less productive equipment. Therefore, any procurement requirement for state-of-the-art machine tools, whether these tools are intended to be used alone or as part of a flexible manufacturing system, CAD/CAM system, or other automated system, almost invariably leads to new purchases. This is true, for example, of the current arsenal and shipyard modernization programs, which provide for equipment purchases as high as $200 million per facility over a period of 5 to 10 years. Although these procurements are large compared to past years, they commonly entail the purchase of only one or a few identical machines at a time.

Unlike the Army and Navy, the Air Force has a central procurement unit, which facilitates somewhat higher volume purchases. The Air Force procurement office is said to have a tendency to "massage" user requirements to produce conformity among users' specifications.

Despite the Defense Department's interest in promoting production efficiency and the use of state-of-the-art technology, the Committee found that a number of legislative and procedural requirements act as disincentives to new technology development and application by DOD prime contractors. For example, the system of annual congressional appropriations creates uncertainty about the future defense products market, and heightens the financial risk associated with any large investment in new, DOD-oriented manufacturing process technologies. Further, there is little contractor incentive to lower costs through new more efficient machine tools when contracts are negotiated on a "cost plus" or other similar basis (i.e., where profits are based primarily on costs).

MANUFACTURING TECHNOLOGY PROGRAMS

The Department of Defense and the three services have a number of programs designed to promote progress in manufacturing technology. The services' Manufacturing Technology (ManTech) programs concentrate on the validation and application of new process technologies. The Navy's and Air Force's Technology Modernization (TechMod) and the Army's Industrial Productivity Improvement (IPI) programs stress cooperative efforts among defense contractors and their suppliers, encourage incentive agreements not necessarily tied to specific weapons programs, and aim to highlight counterproductive aspects of DOD's procurement process.

The DOD has recently started implementing policies to bring TechMod and IPI under one name, Industrial Modernization Incentives Program (IMIP). As a new program designation, IMIP is as yet unfunded. The DOD budgets for the ManTech, TechMod, and IPI programs appear in Table 13.

Although separately and variously administered by the services, the three ManTech and TechMod (IPI) programs have several common features.

ManTech Programs

The Manufacturing Technology program, dating from the early 1950s, is designed to promote the development and application in defense production of new manufacturing processes previously validated in the laboratory but not yet reduced to economically sound practice. The program concentrates on situations where industry is unable or unwilling to commit private resources, at least on a timely basis, to make technologies available for use in meeting DOD requirements.

TABLE 13 DOD Manufacturing Technology Program Budgets ($ million)

		Request		
		FY 82	FY 83	FY 84
Army	ManTech	93	$50[a]	101
	(Manufacturing Methods & Technology)			
	Industrial Productivity Improvement			
Air Force	ManTech	61.8	66.4	
	TechMod	34.0	38.0	
Navy	ManTech	37.3	49.8	
	TechMod (included in ManTech funding)	6.0	6.0	

[a] A House Appropriations subcommittee first rejected the Army's FY 1983 request for ManTech, then added $50 million—but under R&D rather than procurement programs.

Supported in most cases by procurement funds, ManTech finances little research and development and generally the purchase only of prototype equipment. It aims to define particular technologies to the point at which they are repeatable and reliable, with the expectation that weapons systems manufacturers will then purchase and use them in volume. ManTech projects are non-proprietary; diffusion is, in fact, encouraged by requirements that the contractor make a disclosure of technical findings and implementation results as well as license the processes developed on a non-exclusive basis.

ManTech projects may be awarded to any qualified performer; equipment vendors are informed of DOD plans and encouraged to bid. In practice, however, all three military services have awarded the overwhelming majority of external ManTech projects to prime defense contractors and independent laboratories. Approximately 40 percent of the Army's ManTech budget is spent in-house. Very few awards have been made directly to machine tool companies.

ManTech funds have gravitated to prime contractors for the following reasons:

- Increasingly, DOD policy has placed a premium on the implementation of validated technologies. Evaluations showing higher technological than implementation success rates have reinforced this policy, as have pressures from Congress and elsewhere. Not only is it the conviction of responsible DOD officials that technology "pull" efforts are more effective than technology "push" efforts, but it is also the prevailing opinion within DOD that prime contractors are generally disinclined to adopt novel production equipment with which they are not very familiar. In these circumstances, reliance on prime contractors encourages the application of ManTech results, though often by sacrificing widespread diffusion. The original contractor is frequently the only user.[4]
- ManTech pays only part of the costs of developing and demonstrating new technologies, usually excluding the costs of prior research, development, and capital equipment. This narrow support is usually attractive only to companies that are accustomed to investing heavily in R&D or are able to bear the prior capital equipment costs. U.S. machine tool companies in general fit neither of these categories.
- Prime contractors and laboratories and consulting organizations dependent upon DOD business have invested heavily in an institutional capability to compete successfully in the defense market. In many cases, this investment includes personnel expert in anticipating ManTech requirements and marketing proposals. For such companies, it is estimated that the cost of developing a proposal for a $300,000 ManTech contract is in the range of $10,000 to $15,000. For those not accustomed to competing in this market, the cost may be two to four times as great and, therefore, prohibitive.

- The regulatory and other disincentives to machine tool company participation in defense procurement apply with equal force to the ManTech program.

Equipment suppliers can and do participate indirectly in ManTech projects as subcontractors and advisors. For example, a current Department of the Army project to disseminate FMS technology has recently resulted in the completion of a large study detailing the economic and technological potential of flexible manufacturing systems. This project, which is designed to overcome a perceived lack of information among machine tool users about the potential of FMS technology, is being carried out through a consortium that includes several machine tool builders.

ManTech supports technologies applicable to the production of a single weapons system or even component, but program guidelines favor the support of generic technologies that may be used in the manufacture of different types of defense materiel. Such technologies are not limited to metal processing, material handling, composites production, and automation, but encompass a wide range of objectives including chemical processing, electronics packaging, energy conservation, and safety and health. Table 14 lists the technological areas receiving greatest emphasis in each of the ManTech programs. Thus, the relatively limited funds committed to the Manufacturing Technology programs as shown in the table are spread among a relatively large number of manufacturing technologies.

The DOD's ManTech programs use conventional procurement terms and procedures. Contracts are usually competitive and negotiated on a fixed price or cost plus basis. In some cases, incentive awards are made for superior performance.

The lead times for ManTech projects do not vary significantly from those for ordinary purchases. A decision to pursue a technology may precede a request for proposal (RFP) by as much as three to five years, and a few months to a year may elapse between the advertisement of an RFP and the contract award.

These long lead times for ManTech contracts seem self-defeating, in view of the program's purpose of promoting advanced technology. Like other parts of the DOD budget, ManTech budgets must be assembled at least two years in advance of contract awards. This means that DOD substantially lags the private sector in its ability to promote rapidly changing manufacturing technology.

TABLE 14
DOD ManTech Programs--Technological Thrust Areas

Army

Metals (including powder metallurgy)
Electronics
Optics
Chemical processing
Pollution control
Testing
Energy conservation
Safety and health
Materials handling
Packaging
Automation
Nonmetals (including composites)

Air Force

Machining
Powder metallurgy
Composites production
Electronics packaging
Flexible automated batch manufacturing
Critical materials
ICAM architecture/applications
Repair operations
Electronic power devices

Navy

Aircraft and related systems:
    Airframe assembly automation
    Materials technology for propulsion
    Avionics, test and evaluation
Ships, shipbuilding and related systems:
    Shipbuilding automation
    Large combat systems structures
    (e.g. gun mounts)
    Hull
    Outfitting and furnishing
    Computer-aided ships engineering
    Mechanical subsystems
    Electrical subsystems
    Auxiliary subsystems
    Shipyard services

Electronic components:
    Microwave devices
    VHSIC
    Electro-optics/fiber optics
    Solid state technology
    Printed circuit technology
    Materials
Logistics:
    Parts-on-demand technology
R&D:
    Flexible manufacturing
      systems
    Welding technology

Source: Department of Defense.

Any effort to increase the direct participation of equipment vendors in ManTech programs must take into account not only the peculiarities of these programs, described above, but also the level and uncertainty of current ManTech funding. In particular, the stability and continued growth of the ManTech program appears in jeopardy as a result of an unexpected congressional action with respect to the Army FY 1983 appropriation. On the initiative of a House Appropriations subcommittee, Congress reduced the Army's request by 60 percent and converted the remaining $50 million from procurement to R&D funds. This action reduces the Army's flexibility in obligating the remaining funds, jeopardizes ongoing projects, and threatens to transfer the program to an administrative environment less sensitive to the requirements of applying and diffusing new technology.

TechMod Programs

The Technology Modernization program and its Army version, the Industrial Productivity Improvement (IPI) program, originated with the F-16 aircraft program in the late 1970s. It is weapons-system-based and plant-based rather than project-based and technology-specific. TechMod/IPI funds the validation of advanced manufacturing technologies in return for a contractor's commitment to make significant capital investments in modernization of equipment producing a particular weapons system in a particular facility. Although its purpose is ordinarily to reduce costs, it may also be used to increase surge capacity or improve product quality and performance.

A TechMod/IPI project may be initiated either by DOD acquisitions personnel or by a contractor. A typical TechMod contract incorporates three phases, which may be negotiated separately. In the first phase, DOD supports a top-down, wall-to-wall analysis of the contractor's production facility. In the second phase, DOD supports the advanced development of identified technologies and the design of plant improvements. Finally, the contractor undertakes to purchase and install the new equipment.

Although it originated independently, TechMod can be and has been viewed as a means of ensuring the implementation of ManTech project results or of promoting other advances in the state of the art. Frequently, however, TechMod results in the adoption of off-the-shelf though technologically advanced equipment. There is a danger that ManTech opportunities are identified too late in the procurement cycle to incorporate them in ongoing weapons programs or are judged to be too long-term and to entail too high a risk to justify immediate adoption.

TechMod contracts are exclusively with weapons system producers, although in the F-16 program and other cases they have been extended through agreements between prime contractor and subcontractor to second-tier component manufacturers. DOD policy encourages this "pyramiding," out of the realization that subcontracted component systems often represent more than half of the cost of a weapons system and out of concern that second- and third-tier suppliers are frequently fragmented, have poorer access to capital markets, and therefore have greater difficulty than primes or major subcontractors in obtaining capital for investment in modern plant and equipment.

TechMod and IPI offer incentives that are not typical of conventional procurement contracts. For example, to protect the contractor in the event a weapons system contract is unilaterally terminated or stretched out because of insufficient funding, DOD may agree to pay the undepreciated value of the equipment purchases by the contractor. Secondly, DOD may agree to a formula for sharing with the contractor the savings resulting from productivity gains. In these cases, the contract stipulates investment commitments for each fiscal year of the contract and targets (though does not guarantee) return on that investment for the contractor. Finally, TechMod contracts frequently use the more conventional device of incentive awards for contractor performance.

The general aim of these and other measures utilized under the aegis of TechMod/IPI is to provide incentives for contractor investments through greatly increased returns on investments and by indemnification of investments in the event of cancellation of the procurement programs for which the investments are made. Government and industry contract specialists have faced several problems that have precluded greater use of these concepts. Where there is more than one product and more than one government buying office with work in a facility, it is difficult to determine which office or which contract should be the vehicle for the special investment agreement. In addition, it is difficult to measure actual savings resulting from new equipment or facilities and to divide the savings between the government and the manufacturer. Also, the government has had some difficulty in providing indemnification against program cancellation or stretchout because of existing rules that govern the contracting process. Efforts to overcome these problems could have a significant impact on requirements for new more efficient machine tools.

A 1980 report of the Air Force Systems Command suggested that "if technology modernization can work for aerospace, it can work for other critical civil/military industries, such as electronics, machine tools, and basic materials."[5] A TechMod program for equipment vendors conceivably could be carried out through prime contractors or directly in connection with DOD procurement. In either case, however, the question arises whether any prime contractor or DOD agency represents a large enough market to justify participation on either side. Machine tool companies do not have dedicated facilities, and defense-related purchases are commonly in small lots. In comparison with ManTech, moreover, contractor participation in TechMod programs requires an even more sophisticated marketing capability, since the contract terms are more complex and the financial commitments greater.

MACHINE TOOL SUPPLIERS' PERSPECTIVE ON THE DEFENSE PROCUREMENT PROCESS

The following analysis of the machine tool suppliers' perspective on defense procurement is based on the views of a range of machine tool companies doing business directly and indirectly with the government. Those interviewed were asked to compare their experiences selling (1) directly to the government, (2) to prime contractors, and (3) to non-defense businesses. They were also asked to cite specific examples of problems and successes. Our field research revealed that, in general, machine tool companies find it more difficult to work directly with the government than with prime contractors or civilian customers.

In the last few years, seeking government contracts (from arsenals, national laboratories, etc.) has been a relatively low priority for most machine tool companies. Direct government contracts were typically 5–10 percent of sales. While government business, in general, is not seen as technically more demanding or more risky than business with prime contractors or civilian customers, government contracts are generally perceived as entailing greater administrative difficulties.

On the other hand, doing business with prime contractors was seen as comparable to civilian business. Lead times between the request for a bid and the contract award were substantially shorter than those experienced when dealing directly with the government. Indeed, the prime contractors were viewed by respondents as useful in shielding machine tool companies from the problems of direct government negotiations.

Dealing With the Government

While not every machine tool company interviewed had concerns about direct government business, a substantial number agreed on the types of administrative procedures in technical specification that tend to discourage machine tool builders. The administrative problems are:

- excessive paperwork
- long lead times
- variation and unpredictability in lead times
- lack of understanding of government procedures and reviews

The problems in technical specifications are:

- lack of understanding of manufacturing at some government installations
- inadequate consultation with the industry before and during the contracting process
- inappropriate specifications, which often result in outmoded or unnecessarily expensive machinery

These sets of problems are seen by suppliers as reasons to avoid dealing directly with the government, especially during periods of high order backlogs. According to these companies, such impediments result in increased costs and impaired quality for the government customers.

The companies claim that administrative problems add delays, uncertainties, and extra costs to the system; excessive paperwork adds unproductive, administrative time for machine tool companies. They argue that this is especially true for the excessive detail, compared with civilian work, with which many requests for proposals are drafted.

More serious problems are found in the contracting procedure itself. Lead times of 12–18 months are common in direct government business, compared with 3–6 months with primes and civilian customers. These long lead times add uncertainties and place a premium on continuity at the companies. They also mean that, as machine tool firms' own backlogs are being worked down, government business is not a viable, short-term alternative.

The greatest administrative problems appear to result from variations in processing time combined with lack of documentation of review procedures. As delays occur, companies that do not know the sequence to be undertaken on a bid have difficulty locating and resolving the source of the bottleneck. One company proposed, as a model for the DOD, the system recently installed at the General Services Administration (GSA). If certain higher-level reviews are not completed in 20 days at GSA, some purchases can be assumed approved, and paperwork moves to the next stage in the process.

The long, complex contracting process tends to favor two types of companies: (1) large companies with multiple government contracts, who can spread the costs of bidding and managing government contracts over a number of jobs and develop long-term relationships with DOD, and (2) certain small firms that are dedicated to obtaining government contracts and whose top management have special expertise in this area. Small companies with few government contracts are at a disadvantage in bidding because they lack the resources, specialized personnel, knowledge of the process, and close relationships to perform well in the bidding process. Skills in contracting, however, do not necessarily coincide with the ability to commercialize and promote the advanced manufacturing technologies in which DOD appears to have the greatest interest.

Another concern of machine tool suppliers lies in the area of technical specifications. Sometimes the specifications do not reflect up-to-date manufacturing technology. Machine tool builders believe that consultation with the industry before machines are specified is inadequate. One cited an example of a government specification for 11 4-spindle, 5-axis machines that would cost about $1 million each, when an already available 4-axis machine for $150,000 could do much of the work required. Previous consultation might have reduced the number of 5-axis models.

Suppliers also believe that custom machines are specified to an unnecessary extent. While the government needs some custom machines, specifications for custom designs can also be used to influence which companies are likely to win the bids. Sometimes the specifications combine the features of a number of manufacturers, which raises costs without affecting performance significantly. One company reported that they had built custom machines for a government contract which were no more effective than their standard product, but which cost two to five times the standard costs. Custom designs may also require the diversion of scarce management and engineering time to machines that will not be useful to other machine tool customers.

Finally, the companies claimed that the initial specifications are seldom updated, and very little technical communication is allowed. With the long lead times involved in contracting and rapidly changing technology, the government can end up with obsolete equipment. One company cited a contract for a computerized design system that was specified in 1978 but not awarded until 1981. By the time the award was granted, computer-aided design (CAD) technology had improved dramatically. However, since the specification was never updated, the company was required to deliver obsolete equipment.

In another example, a company became aware that a specification for a group of machines costing over $10 million had been written for a job. The firm believed that a group costing less than $6 million and a new manufacturing approach could have solved the problem, but because the contract was already written, the government customer would not consider a new, less costly approach.

The solution to these problems is, to the extent possible, to specify the parts to be produced and let companies bid machines to fulfill the job. This approach takes advantage of the machine tool companies' expertise in manufacturing and should result in expensive custom machines being bid only when absolutely necessary. Government personnel would then have to devise a scoring system to judge, based on cost and design, the most effective, lowest-cost alternative and select that manufacturer.

The difficulties that many machine tool firms have in dealing with the government are summarized in recent congressional testimony by Richard P.Bodine, the president of one of them:

In the 1970's, our industry—like all others— was buried in burdening government regulation. We were made aware during Vietnam that if we won a competitive fixed price contract—even in our high-risk business—the contract price was not firm. We were subject to government audit, to insure we were not "too profitable." We were told that "advertising expense" was not permitted, since no one had to advertise to get a government contract. We were also prohibited from paying dealer commissions on government sales—even though our arrangements with our dealers/ representatives had existed for years. We are legally obligated to provide support as required to the auditors from our limited staff—without compensation. We are too small to handle this kind of intervention.

In short order, we were faced with EEOC, OSHA, EPA regulations, Affirmative Action, ERISA, and many others. Most are mandatory—some are voluntary, but mandatory if you wish to do government business. In 1972, we made a corporate decision to avoid any government regulations we legally could avoid. As a result, we are totally dedicated to civilian customers. It is no longer economically practical for us to bid for or accept a government order.[6]

An Important Counter-Example

A recent example of a procurement machine tool companies cite as exemplary, from both administrative and technical viewpoints, was the Watervliet arsenal purchase of an FMS. Although the project is not yet complete, members of the machine tool industry believe it will demonstrate excellent results. Watervliet recently awarded White-Consolidated Industries a $15.3 million contract for a fully automated, flexible manufacturing system for howitzer and gun tubes. The FMS incorporated as major components a number of horizontal machining centers and vertical turning machines and included sophisticated coordinate measuring machines and an integrated material handling system under the control of a large minicomputer. The committee interviewed not only the winner, but also a loser, of the contract, and both praised the method of purchase as a model for others.

The Watervliet procurement process had a number of distinguishing features. First, the DOD personnel involved understood manufacturing and contracting very well, and in general specified the outlines of an FMS as modern as any ever built. Second, within the general guidelines, companies were allowed to design their own systems. As a result, the final designs offered a variety of approaches, bringing out what each company considered the best system.

A scorecard system was devised beforehand, announced to the companies, and used to evaluate the bid. Points were awarded for efficiency of the system, flexibility of software, and accuracy of tools, as well as for price. The company with the highest point total was selected.

Another critical feature was that, under strict controls, the government allowed technical communication between the bidding companies and Watervliet personnel. This procedure allowed companies to gain vital information on the requirements of the system, while protecting the process from abuse.

The major differences, then, between this Watervliet and other DOD manufacturing technology purchases were that (1) the most modern system was a specified objective, (2) companies had the freedom to design their own systems without having to meet detailed specifications that would limit their options, and (3) technical communication was kept open. Instead of a potentially costly, out-of-date system, the Defense Department will be receiving a modern system after a strongly contested bidding process. Although the bidding process took longer than normal, and although most companies said they did not expect to profit from the sale to Watervliet, these firms were enthusiastic about participating because of possible commercial spinoffs.

Experiences such as the Watervliet project can do much to change the generally negative perception that machine tool companies have of dealing directly with the government. The Department of Defense can still have powerful leverage within the industry through making a market for new technology, as it did in the case of numerical control. Machine tool companies can be strongly motivated by government procedures that take the trends in the industry and plans of the firms into account. At the same time, better cooperation will make it possible for the Defense Department to obtain better manufacturing technology, more enthusiastic company participation, and, to the extent that standard machines replace custom designs, lower costs.

PERCEPTIONS OF THE U.S. MACHINE TOOL INDUSTRY: THE PRIME CONTRACTORS' VIEWPOINT

This subchapter describes the role of the prime defense contractor as the user and the developer of manufacturing technology. After some initial observations, the subchapter takes up several issues that are key to the prime contractor's role: in-house machine tool-making capability and performance, contracting procedures, technology flow, experience with foreign suppliers.

The Committee found that several issues are key to the contractor-supplier relationship: the in-house machine tool-making capability of prime contractors, the contracting procedures between primes and their suppliers, the sources of new technology in materials processing and handling, and the experiences prime contractors have had with foreign vs. domestic suppliers.

In-House Machine Tool Capability and Performance

None of the firms interviewed for this report had machine tool fabrication divisions or met all machine tool needs in-house. Complex machine tool design, however, was carried out as a function of manufacturing research. In such cases, the firm's research division might construct a prototype, turn it over to the firm's facilities division for testing and refinement, and then contract with a machine tool manufacturer for final production. A conventional arrangement of this kind would involve the machine tool builder as a licensee to patents held by the firm. Respondents interviewed for this report cited at least two examples of such a procedure, involving a dry ice pellet blaster and a tape-laying machine.

This management of in-house facilities appeared to occur even at large companies whose product lines include some machine tool components. At one such company, in-house machinery has been used to build some of its (non-defense) products. A spokesman for the company stated, however, that his firm was "not in the machine tool business." An aerospace contractor was more emphatic: "There is no way the primes can compete with machine tool builders on their own turf."

On the other hand, virtually all prime contractors maintained some machine tool capability. Various reasons were given for this. At a minimum, a machining capability was required to adapt existing machine tools for special jobs and configurations. In isolated instances, a company with a proprietary interest in a specific manufacturing technology might prefer to construct the machines utilizing this technology in-house rather than allow it to become widely known.

Thus, machine tool fabrication by prime contractors typically involves the shaping of cutting equipment, and some blank and mill grinding. Some respondents said that their firms regularly manufactured their own machine tools when the mechanics were simple and peculiar characteristics were required. A minority of (usually large) firms, however, regularly fabricate major machine tool assemblies. These situations have usually taken place where extremely specialized manufacturing processes are required or when proprietary information is involved.

Contracting Procedures

Contracting procedures between the primes and their machine tool suppliers vary according to the cost of the equipment, the extent of new technology to be incorporated, specialized requirements such as short lead times, and the way the corporation itself organizes its research, purchasing, and production functions. The purchase of any major machine or machine system, however, usually entails specification-writing, coordination with the using activity, the bidding process, and "run-out" or on-site testing.

Where especially sophisticated or new technology is a critical element of the machine in question, a prime's manufacturing research division will play a role at most of these stages. It may already have built a prototype of the machine in question. It will, at any rate, help draft specifications around the requirements of the using activity. Divisions "signing off" on the equipment purchase could include the using activity, facilities planning division, and maintenance. One aerospace firm reported that this procedure (using manufacturing research personnel to manage a new manufacturing technology purchase) covered approximately 25–30 percent of all machine tool purchases in the average year.

Firms normally attempt to draft specifications so that several suppliers might be capable of bidding. The bidding process, however, is not handled uniformly among all defense contractors. Although many contractors use a "three-bid" or "four-bid" procedure, some are known to open the bidding to a wider range of suppliers. And one respondent stated that he seeks out smaller suppliers, often on a sole-source basis, that he knows to be as competitive as the larger foreign and domestic firms.

If the machine in question employs significant new technology, the prime must often work closely with the machine tool builder to encourage it to embark on such a project. This subject is covered in more detail, below. It is important, because the responsiveness of the machine tool builder to advances in the state of the art has become an increasingly important factor in the industry's competitiveness.

Purchase contracts generally provide for on-site testing, or "run-out," especially where the state of the art is being pushed.

Technology Flow

Research and development budgets in the machine tool sector have historically been modest, in absolute figures and as a percentage of sales. In 1981, machine tool industry R&D stood at 4.2 percent of industry sales, and even this figure needs to be qualified by the observation that much of that figure represents development, as opposed to research, spending.

Many prime contractors, on the other hand, appear to place considerable emphasis on the application of emerging technologies. Research divisions at leading defense contractors work with their manufacturing divisions and purchasing departments at the various stages of the equipment procurement and testing process, whenever new technology is involved. Such firms strive to maintain strong ties with universities, where the bulk of the nation's basic research is performed. A recent report on the aerospace industry, for example, concludes that aircraft and engine manufacturers "presently carry out an extensive and multifaceted university interface, covering virtually every form of industry/university relationship."[7]

Examples of research performed, or contracted for, by prime contractors include titanium shaping, polymer fiber breakage, honeycomb metal forming, heat shield forming, and tape laying—i.e., subjects germane to machine tool characteristics and specifications. Several prime contractors commit more in certain years to manufacturing research alone than does any single domestic machine tool firm.

These differences between prime contractors and machine tool makers in their R&D budgets also influence the way technology has developed in the machine tool industry. Although there are exceptions, technology flows in manufacturing processes have over the last two decades generally originated from outside the machine tool industry: from government, private, and university laboratories; from prime contractors; and in some cases from foreign manufacturers.

In its interviews, the Committee found that a number of features of the prime/DOD/machine tool industry relationship have helped to inhibit the development of a steady source of technology flows from within the U.S. machine tool industry. The most prominent features are the following:

- Industry structure and practices. The relatively small size of the average U.S. machine tool firms, and the peculiar economics of machine tool sales that affect even the largest firms, have limited the amount of useful basic research that can be performed or that U.S. machine tool firms have been willing to finance in-house. Machine tool construction using new technology, therefore, has tended to be "customer-driven" and not originated by the industry.
- Comparatively slow domestic market for new production technology. The lack of a substantial domestic market, until recently, for the latest manufacturing technology has affected progress within the machine tool industry. This issue is discussed below in further detail.
- Difficulties of direct DOD-supplier contacts. As this report describes elsewhere (see "Machine Tool Suppliers' Perspectives on the Defense Procurement Process", above), many machine tool builders do not take advantage of government research contracts. The bulk of the DOD-sponsored research in manufacturing technology, for example, is performed by prime contractors rather than machine tool builders.
- Primes' advantage regarding unique machine tools. Because of their size, large prime contractors are in a better position than most machine tool companies to develop and construct the sophisticated, one-of-a-kind machine tools that are often used for building critical parts of advanced weapons systems. Larger firms can spread the research and development costs of these tools through such mechanisms as Independent Research and

Development (IR&D) allocations on their government contracts, which as a practical matter are unavailable to most machine tool builders.

Prime contractors interviewed for this report stated that although U.S. machine tool builders have kept abreast of technological developments in some areas, they fall short in others. According to these respondents, U.S. machine tool firms are behind the state of the art in applying flexible manufacturing systems and in some applications of computer technology. This perception is disputed by leading U.S. machine tool firms, which claim that U.S. FMS technology is at least equivalent to Japanese technology.[8] Their position has some support in the literature,[9] including a recent survey of relative technological positions by Japan's Ministry of International Trade and Industry (MITI).[10]

To the extent that the prime contractors' judgment is true beyond the survey sample, however, it is especially ominous, inasmuch as the areas they cite—FMS technology and some applications of computer technology—are where some of the most significant gains are being made in manufacturing technology. If this judgment is not accurate presently, it could become accurate soon, because the Japanese government is spending at least $60 million to improve commercial FMS technology.[11] Three national research institutes and 20 companies are participating in this program.[12]

There does appear, at any rate, to be a perception among machine tool users that the U.S. products are generally inferior, whether or not the perception is warranted. The Committee did not identify the extent to which this perception is the result of marketing vs. technological factors.

The Committee believes it is important, however, to compare these perceptions with the observation that foreign manufacturers that use machine tools, especially the Japanese, appear to have made significant investments in modern machine tool technology before their U.S. counterparts did. The reasons for this advantage could include such diverse factors as more enlightened labor-management relations in Japan, built-in disincentives to manufacturing efficiency in the United States because of "cost-plus" provisions in DOD contracts, and the relative effects of U.S. vs. Japanese incentives for capital investment. It is commonly agreed, however, that the Japanese suppliers brought to the U.S. market in the mid to late 1970s more experience in some sophisticated categories of machine tools than U.S. tool builders. With regard to flexible manufacturing systems, for example, observers point out that Japanese machine tool builders had a head start in commercialization, because U.S. machine tool users—in contrast to Japanese users—were slow to pick up on the concept.

This observation accords with the Committee's experience that, in making machine tool purchases, U.S. firms have until recently had a tendency to "replace" rather than "upgrade." The decision to purchase has often involved low-level or uninformed decision-making (e.g., by foremen or purchasing officers). This has colored the perception prime contractors have had regarding the responsiveness and reliability of U.S. vs. foreign suppliers of machine tools.

Experience With Foreign Suppliers

Although some prime contractors strongly prefer to buy from U.S. suppliers, all interviewed respondents stated that they made substantial machine tool purchases from foreign companies. The most commonly cited disadvantages ascribed to U.S. suppliers were these:

- <u>Delivery times</u>. As this report examines, machine tool imports have tended to climb during those years when U.S. suppliers were accumulating large backlogs. In the case of the latest surge in imports, which took place during 1976–80, U.S. buyers found that the overseas supplier could deliver an order several months before its U.S. competitor.
- <u>Responsiveness to user requirements</u>. Most machine tool users that responded to the Committee's surveys believed that foreign manufacturers were more responsive to user requirements, especially where state-of-the-art advances were involved. Some named specific instances where U.S. suppliers had turned down opportunities to bid on projects incorporating new technology; these bids had subsequently been picked up by foreign firms. In one instance an aerospace firm decided on specifications for a large, multiple-spindle profiler with automatic tool-changing and pre-setting capability. It received seven bids; only one U.S. firm was among the bidders. "The U.S. machine tool industry has kind of left us," the

aerospace company's General Manager for Manufacturing Operations remarked.
- After-sales service. The majority of interviewees also faulted follow-on service standards at the U.S. machine tool firms. In some cases, poor follow-on service appeared to result from the "conglomeratization" of the supplier. Respondents who brought up this point surmised that where service had once been provided by distributors, whose prime responsibility lay in sales and service, it was now done directly by the supplier-conglomerate. Follow-on service thus became a lower corporate priority and suffered accordingly.
- Reliability. The reliability of U.S. machine tools came in for some of the strongest criticism. As the head of manufacturing research at an aerospace firm put it, "The Japanese are more likely to give you a product that will run the first time: U.S. manufacturers usually give you a longer lead time, and the reliability of their machines is not the greatest." Another, similarly placed corporate officer likened the situation to the U.S. auto industry, which he described as outclassed by foreign products that offer better reliability and are more responsive to consumer demands.

The Committee acquired anecdotal but nonetheless persuasive evidence to the effect that prime manufacturers are seeing improvements in the competitiveness of U.S. machine tool builders. The Petition of the NMTBA for relief under Section 232 of the Trade Expansion Act also describes in some detail the "self help" steps being taken by the industry.[13]

Together, these suggest that U.S. machine tool builders are aware of changes that must be made in order to remain competitive. As previous sections of this report suggest, these changes will be constrained by financial considerations, and by the difficulties that U.S. suppliers have had in dealing with the government. The following subchapter describes aspects of U.S. legislation that have influenced and will continue to influence the purchase of domestically produced machine tools during this transitional phase in the industry.

DOMESTIC LEGISLATION AFFECTING THE PURCHASE OF U.S. PRODUCED MACHINE TOOLS

"Buy America" and Other Preferences

As a general rule, U.S. government policy favors domestic over foreign suppliers. The Buy America Act, for example, requires that materials and supplies purchased directly by the U.S. government be composed substantially of domestic products. The Air Force's Buy United States Here (BUSH) program has established procedures so that U.S. products will receive higher priority in procurement among overseas procurement agencies. The Small Business Act gives certain preferences to metalworking machinery producers having 500 or fewer employees; this covers all but 3 percent of U.S. machine tool firms. Several Executive Orders provide incentives for firms performing contracts and planning new production facilities in labor surplus areas; these areas presently include the home territory of many machine tool companies.

This report finds no evidence that such incentives have had a measurable effect on U.S. machine tool purchases by defense contractors. The Buy America Act does not apply to machines purchased for a contractor's own use; it does not apply to the software used to run automated machinery; nor does it apply to purchases from NATO countries, Switzerland, Australia, Israel, or Egypt, where the United States has Memoranda of Understanding (MOUs) waiving the Buy America requirements that might otherwise attach to the purchase of machine tools. The Small Business Act preferences have apparently not served to bring smaller U.S. firms up to the competitive standards of foreign market participants, and at any rate do not reach the firms that account for a very large share of the sales of domestically produced machine tools. The labor surplus area programs do not affect the price competitiveness of the finished product.

Moreover, free trade policies embodied in the Trade Agreements Act of 1979, and in a number of reciprocal international agreements (including the MOUs referred to above), encourage foreign firms to seek host government contracts and provide for the waiver of domestic preferences.

Offset agreements with foreign governments also divert purchases, including machine tool purchases, to foreign soil. These agreements are intended to assist in financing foreign military sales, by providing that the

THE DEPARTMENT OF DEFENSE, PRIME CONTRACTORS, AND THE MACHINE TOOL INDUSTRY: RELATIONSHIPS THAT AFFECT INDUSTRY STRUCTURE

U.S. prime contractor purchase certain components or assembly equipment from the receiving government as a condition to the contract. A recent Treasury Department report estimates that between 1975 and 1981, 26 of the largest electronics and aerospace firms provided foreign governments with offsets totaling $9.5 billion, in return for $15.2 billion in foreign military sales.[14]

Interviews for this report confirmed these conclusions as to the ineffectiveness of this legislation in encouraging domestic machine tool purchases. In virtually every case, Buy America preferences did not, as a practical matter, stand in the way of users that preferred the foreign machine tool over a similar, U.S.-made version.

Legislation recently introduced in the House of Representatives (but not enacted) addresses some of the concerns raised in this report. The bill, HR 2782, would set up a 3-year, $1.8 billion program of modernization and expansion loans for defense-related small and medium businesses. It would also establish training programs throughout the country to help reduce shortages in certain, largely vocational, labor skills. Finally, the bill would provide for grants to colleges and universities to purchase and install modern scientific and engineering equipment.

A committee report accompanying the bill points out that the legislation has as its intent "increasing productivity, improving product quality, and lessening import dependence." The legislative history of the bill indicates that it was drafted with the machine tool industry, among others, in mind.

Machine Tool Stockpiles

Under the Defense Industrial Reserve Act (Public Law 93–155), the government is authorized to procure and manage a stockpile of weapons parts and also of manufacturing equipment such as machine tools. DOD's stockpiled machine tool (metal-cutting and -forming) inventory consists of two categories: (1) the General Reserve, which is centrally managed by the Defense Logistics Agency, and (2) various idle packages for mobilization, which are managed by each of the three services.

(1) As of July 1983, the General Reserve had an inventory of 12,286 machine tools, which were valued at

$334 million. However, the average age of these tools is 29 years, with only 2.1 percent of the metal-forming tools less than 10 years old, and 1.2 percent of the metal-cutting tools less than 10 years old.

Longstanding DOD policy has aimed at replacing 5 percent of this inventory each year; but because of the lack of funds, this goal has not been met. In 1981, the Defense Science Board recommended a one-time, 25 percent replacement and a 5 percent replacement thereafter; this recommendation, however, has not been implemented.

(2) The Idle Packages for Mobilization numbered 13,489 machine tools as of July 1983, with an inventory value of $382 million. Similar to the inventory in the General Reserve, however, the average age for the Idle Packages inventory is 28–29 years.

The 25,775 DOD-owned machine tools currently in storage is down from an estimated 32,000 tools in 1981. Because of a lack of funds, this significant reduction in inventory has not been matched by an increase in the number of new tools.

The Committee believes that the whole concept of long-term stockpiling of machine tools by DOD needs to be examined carefully. A recent Army report, for example, asserted that use of the stockpile to provide machine tools for M1/M60 tank production would "cost a great deal of money [in machine tool rehabilitation] and would not improve manufacturing methods above those used for the last fifty years."[15] Thus far, the stockpile concept has tended to discourage technological advance while running up substantial carrying charges for the taxpayer.

INDUSTRIAL BASE RESPONSIVENESS

In its effort to analyze this country's ability to respond to wartime production requirements, the DOD regularly publishes mobilization plans for specific weapons systems. The two described here give an indication of the continuing need for a responsive machine tool industrial base.

In May 1978, the Army published its study of surge and mobilization requirements for the M109A2, 155-mm self-propelled howitzer. It found that the cannon and spare tubes for the M109A2 were critical pacing items. This problem was highlighted by the lead times for the construction of industrial plant and equipment; these lead times did not match the accelerated production requirements of the surge scenarios in the study. The study concluded, "the long lead time required by the tooling industry to produce industrial plant and equipment is a critical problem area pointed out by this study and should be of interest to DOD."[16]

In June 1982, the Army published an industrial preparedness study for the M1/M60 tank systems. Among its purposes was (1) analyzing current production capabilities, (2) identifying critical and pacing purchased items where the vendor could not meet mobilization requirements, and (3) identifying machine tools, production equipment, and tooling required to meet mobilization planning.

That study, which cost more than $900,000 to complete, found that the new machine tools and production equipment required for mobilization are long lead items, not available off the shelf. It concluded that "to meet mobilization requirements and update manufacturing methods will require 200 new machine tools and an additional 200 pieces of special equipment with a producible lead time of 18 to 24 months."[17] The study doubted that the American machine tool industry could accomplish this task in today's industrial environment.

CONCLUSIONS

In the course of its interviews and surveys, the Committee was struck by several features of the DOD-prime contractor-supplier relationship that have served as disincentives to modernization in the U.S. machine tool industry. These can be summarized as follows:

- Contracting procedures. The Committee cannot avoid the conclusion that the complexity of the Defense Acquisition Regulations is at least part of the reason why the U.S. machine tool industry has generally avoided direct DOD relationships.
- Market characteristics. The apparent slowness of the machine tool builders' domestic market, which includes prime contractors, to adopt modern production technology on a widespread scale has also affected the competitive status of U.S. suppliers.
- Prime contractors as buffers between DOD and suppliers. While prime contractors generally shield machine tool companies from having to deal directly with the government, they also strongly filter government programs. Machine tool companies interviewed for this

report expressed almost no awareness of the TechMod or ManTech programs. While money can be used from these programs to help purchase tools, no machine tool company interviewed knew if it had made sales supported by these programs.

The reasons for this relative disadvantage of machine tool builders in the defense contracting business cannot be ascribed to any single feature of procurement practices or industry structure. This report has described how delays, regulatory requirements, and lack of information have served to the disadvantage of the traditional machine tool builder.

The Committee notes some instances where progress might be made. Experiences such as the Watervliet project, and the streamlined review procedures at the General Accounting Office, can contribute to reversing the generally negative perception that machine tool companies have of dealing directly with the government.

- Research and development. Chapter 2 of this report pointed out the low levels of R&D spending in the U.S. machine tool industry. This chapter has identified four aspects of the DOD/prime contractor/supplier relationship that have helped perpetuate this situation.
- Industrial base responsiveness, and stockpiling. An analysis of machine tool requirements for producing major weapons systems at surge/mobilization levels is clearly beyond the scope of this study. However, the work that has been done confirms (1) that peacetime levels of machine tool inventories are not sufficient alone to meet surge and mobilization needs, (2) that it is unlikely that sufficient congressional appropriations will be passed in the near future to bring the DOD's machine tool inventory up to reasonable standards of either modernity or surge/mobilization readiness, and (3) that current stockpiling practices have resulted in the maintenance of old and at least partly obsolescent equipment.

The Committee points out, however, that the Department of Defense can exert a powerful influence within the machine tool industry by making a market for new technology, as it did in the case of numerical controls. This would require changes not only in stockpiling procedures, but also in the patterns of manufacturing R&D and in procurement procedures which sometimes leave the government with expensive, obsolete equipment.

The picture of DOD-prime-supplier relationships that emerges from interviews and the published literature is one of a traditional structure that presently does not serve either the government or the machine tool industry particularly well. Progress in improving these relationships has been slow and isolated, which has contributed to deficiencies in the competitiveness of the domestic industry.

NOTES

1. U.S. Department of Commerce.

2. David Henry, "Defense Spending: A Growth Market for Industry," in <u>1983 U.S. Industrial Outlook</u>.

3. In connection with the NMTBA's Petition under the National Security Clause, Section 232 of the Trade Expansion Act of 1962 (19 U.S.C., sec. 1962) for Adjustment of Imports of Machine Tools ("Petition").

4. See, for example, Committee on Computer-Aided Manufacturing, <u>Innovation and Transfer of U.S. Air Force Manufacturing Technology</u>, National Academy Press, 1981.

5. AFSC Headquarters, <u>Payoff 80</u>, p. 25.

6. Testimony of Richard P.Bodine, President, The Bodine Corporation, before the Subcommittee on Economic Stabilization, Committee on Banking, Finance and Urban Affairs, U.S. House of Representatives, May 19, 1981.

7. Aerospace Industries Association, <u>Meeting Technology and Manpower Needs Through the Industry/University Interface</u>, 1983.

8. Testimony of Richard T.Lindgren, president and Chief Executive Officer, Cross & Trecker Corporation, before the International Trade Commission, June 28, 1983, p. 2; Testimony of Michael W.Davis, President, White-Sundstrand Machine Tool Company, before the International Trade Commission, June 28, 1983, p.6.

9. For example, in its May 1983 issue, <u>American Machinist</u> reported on a full-scale FMS built by

Cross & Trecker that became "fully operational [in May] cutting aircraft and missile parts" for Hughes Aircraft Company. The article reports that Hughes' "management mandated that the most modern, state-of-the-art equipment would be provided" and that although "Hughes has no 'buy-American' policy, ...all bidders were U.S. firms" (pp. 109–11).

10. The May 1983 issue of <u>Metalworking Engineering & Marketing</u> reports MITI's conclusions that the product technology achievement level of Japanese machining centers and the production technology achievement level of Japanese package software are substantially inferior to those achieved in the United States. Specifically, MITI concluded that "Japan ['s machining centers are] considerably behind the U.S....in spindle speeds, maximum allowable torque, main motor output and cutting efficiency. ...In precision machinery technology,...Japan is behind the U.S. and West Germany. Japan is also behind the U.S. in design technology, where the U.S. is pouring effort into CAD/CAM." The article suggests that "[t]he reason for the large gap [in machining center technology] is that Japan concentrates on popular general machining centers featuring economy, while the U.S. and West Germany concentrate on special high performance machining centers for aircraft and the like" (pp. 76–83).

11. <u>American Metal Market</u>, July 11, 1983, p. 11A; Statement of Eli S.Lustgarten, Vice President, Paine Webber Mitchell Hutchins before Subcommittee on Economic Stabilization of House Committee on Banking, Finance and Urban Affairs, July 26, 1983, p. 18.

12. <u>American Metal Market</u>, July 11, 1983, p. 11A.

13. "Petition," pp. 221–7; Supplement to Petition, August 30, 1983, pp. 25–40.

14. Department of the Treasury, <u>Survey of Offset Coproduction Requirements</u> (1983).

15. General Dynamics, Land Systems Division, <u>M1/M60 Tank Systems Industrialization Preparedness Mobilization Study, Final Report</u> (June 1982). Vol I, p. 4.

16. Hq U.S. Army Armament Materiel Readiness Command, <u>Industrial Base Responsiveness Study for Howitzer, Medium, Self-Propelled 155mm, M109A2</u> (May, 1978).

17. General Dynamics, op. cit., p. 2.

4

PROBLEM SYNTHESIS AND RECOMMENDATIONS

PROBLEM SYNTHESIS

What is Happening to the U.S. Machine Tool Industry?

The U.S. machine tool industry, once the most productive and technologically advanced in the world, has lost a substantial proportion of its domestic market to foreign imports. U.S. machine tool builders are under considerable pressure from Japanese machine tool products, which some consider to have incorporated superior technology. At a time when a severe recession has eroded and sometimes erased profit margins, rapidly changing manufacturing technology has created new urgency for plant modernization and investment in R&D in the machine tool industry itself. In addition, many customers of U.S. and foreign machine tool builders believe that the Japanese have invested more than the United States has in developing an effective world marketing and servicing network.

Compounding the situation faced by the traditional U.S. machine tool industry is this Committee's observation that the very business of selling stand-alone tools that cut, form, and shape material in production processes has changed radically. As discussed in Chapter 2, manufacturing process improvement needs today are being met by a group of suppliers of computer and systems technologies in addition to the builders of the machine tools themselves.

Because the technology is changing so rapidly and customer needs are increasingly difficult to meet, the problems faced by traditional U.S. machine tool builders are exacerbated. For instance, order-backlog management will not substitute for strengthening efforts to identify and meet customer needs.

Even the more traditional economic forces in this industry are unlikely to reverse the situation. Historically machine tool orders lag the business cycle, but because market penetration by foreign competition seems to be here to stay, the competitive climate faced by U.S. machine tool builders is unlikely to improve even if the current business recovery should prove to be a sustained one.

How Did the Industry Get This Way?

In every advanced industrial country, there are now intense pressures to take a global view of sources of materials, production facilities, and particularly markets, in order to compete successfully. This globalization of business has already taken place in such basic industries as computers, telecommunications, steel, and commercial aircraft. The machine tool industry also appears to be subject to these same forces, which are fed by the more rapid diffusion of technology, changing economies of scale induced by new automated production techniques, lowering of transport and communications costs, and a narrowing of income differences between the United States and other industrialized competitors.

The U.S. machine tool industry is being forced to adjust to these far-reaching developments because its traditional practices are ill-suited to the present day. Unlike their Japanese competitors, most U.S. machine tool builders have managed business cycle swings by accumulating backlogs rather than expanding capacity and marketing. Although the machine tool industry's profitability had been healthy from 1974–1981, its capital investment for modernization has been relatively low. It is losing market share to an industry in a country, Japan, that has lower wage and compensation levels, lower interest rates, and a form of government-industry cooperation that is geared to an "export-or-perish" economy.

The users of machine tools have also influenced the status of the U.S. machine tool industry today. In some machine tool categories, penetration of the U.S. market by foreign firms has been possible because foreign machine tool builders gained important experience with very sophisticated domestic users. With the possible exception of some manufacturers in the U.S. aerospace, farm equipment, and off-road-vehicle industries, there are no U.S. manufacturers with installed processes of the technological sophistication that can be found in West Germany and Japan. The largest U.S. market for machine tools, U.S. automobile manufacturers, has not until recently been a strong articulator of demand for high levels of manufacturing technology.

Although the above paragraphs describe traditional U.S. machine tool builders in general, leaders within that industry have acted and are acting to meet new market realities. This report has shown that the response to new competitive conditions has been widespread and varied. Cost-cutting (including relocation of manufacturing facilities overseas), mergers, joint ventures, diversification into new technologies, more R&D spending, and even a basic reorientation of business strategy have been documented. In addition, as this report describes in Chapter 2, the structure of the machine tool industry is changing significantly. The industry is being augmented by an increasing number of U.S. manufacturers offering products that are becoming an integral part of new manufacturing process technologies.

What are DOD's interests Regarding the U.S. Machine Tool Industry?

This report has identified three levels of DOD interest and concern with regard to the machine tool industry:

1. <u>Access to State-of-the-Art Technology</u>. The DOD is answerable both to its mission of national security, and to interested parties such as the U.S. Congress, for maximizing the reliability, effectiveness, and economy of its equipment and materiel. This requires machine tools and systems of the broadest, latest, and highest capability.
2. <u>Cost-Effective, Expandable Production</u>. The same considerations also require that cost-effective production be readily expandable and sustainable during periods of potential supply-line disruption.
3. <u>Health of the Economy</u>. Because investment in more efficient production—including defense production—is more likely to take place during periods of high levels of economic activity, the DOD is concerned about the health of the economy and of the manufacturing sector.

The Committee found that these DOD concerns and interests will be best satisfied when three conditions are being met:

1. The most appropriate, up-to-date production technology is being widely used in the domestic industrial base, in both prime and second-tier contractors;
2. The use of new technology extends beyond the defense sector, at least to those parts of the civilian sector that might be expected to be diverted to supporting military production during wartime; and
3. The strategic industries that face rapid technological change are also keeping up with the state of the art and maintaining a sound financial position.

While several prime defense contractors are working with leading edge manufacturing technologies, the Committee is concerned that advanced manufacturing technology is not as widely applied in this country as in Japan and Western Europe. While some of the world's best production technology can be purchased in this country, and while delivery times of U.S. machine tools have become more competitive recently, many domestic machine tool users believe that American machine tool firms are not satisfying demands with regard to price and reliability as well as some Japanese suppliers.

What are DOD's Policy Options, Levers, and Constraints?

Although DOD cannot alone galvanize the machine tool industry, the Committee is impressed with the influence that the DOD can have in advancing the development and application of state-of-the-art production technology. Although the size of the direct DOD demand for machine tools is small in volume relative to machine tool sales nationwide, the DOD-induced demand for machine tools is large, and the range of DOD equipment and materiel needs is so wide as to require virtually every form of manufacture in use in the country. Therefore, DOD can, through procurement specifications, affect the standards of production that are used.

In short, DOD is at least partially in a position to ensure that its own interests and concerns vis-à-vis the machine tool industry can be satisfied. It is perhaps the only federal agency so well positioned.

The Committee believes that technological leadership, involving not only the ability to perform state-of-the-art scientific research but also the ability to apply and incorporate it economically into commercial products and processes, will determine the competitive success of the domestic machine tool industry in the global marketplace and will, therefore, be critical to its continuing health. As noted above, this leadership is a function of both the builders and users of machine tools. DOD can do much to stimulate and encourage U.S. builders toward decisive technological leadership in key aspects of manufacturing sciences.

The policy tools available to government range from grants and subsidies, to regulations directly affecting an industry, to other policies designed to provide the conditions that encourage certain desired activities. In the case of the Department of Defense, the most direct influence that can be brought to bear upon the machine tool industry is through procurement. Additionally, DOD can substantially influence the long-terra health of the industry by supporting industry-wide efforts to fill two of its prime needs: better research in manufacturing technology, and a knowledgeable customer for the resulting process technology.

In considering the range of possible actions, the Committee emphasizes that the current situation is not subject to a "quick fix." On the contrary, the only valid solution is one that prepares an already diverse industry for a climate of continuing rapid technological advance and strong foreign competition in domestic and world markets well into the future. Additionally, the financial condition of many machine tool builders militates in favor of a mix of measures having immediate as well as long-term impact. Anything short of a comprehensive package, the Committee believes, could prevent the U.S. machine tool industry from continuing its adjustment to new competitive conditions and strengthen the case for emergency measures in the future.

RECOMMENDATIONS

The following recommendations numbered in order of priority fall into categories of action open to the three key participants: DOD; parties outside DOD's direct jurisdiction but within its power to influence, such as prime contractors and other government agencies; and U.S. machine tool builders.

PROBLEM SYNTHESIS AND RECOMMENDATIONS 89

Recommendations for DOD

1. <u>Modernize the Defense Industrial Base</u>

There is considerable evidence that contracts for advanced weapons systems are often undertaken today using manufacturing technology that is 20 to 30 years old. As a result, not only do these weapons systems have costs that could be avoided, but the country loses opportunities to pioneer in new production technology having a large potential impact on the economy.

To a disturbing extent, the Committee believes, the technology lag in defense contracting reflects shortcomings in the contracting process itself. As explained in Chapter 3, there are too few incentives built into the process to encourage widespread modernization in defense-related production.

One hopeful sign is that DOD has in place programs that could, if given sufficient priority within the Department, make substantial progress. The new Industrial Modernization Incentives Program (IMIP), and DOD's ManTech and TechMod (including the Army's IPI) programs have the potential of speeding the implementation of new manufacturing technology.

Action: DOD should display a greater commitment to the aims of its manufacturing productivity incentives programs. The IMIP (as the successor to the TechMod and IPI programs) and ManTech programs should receive increased and stable funding.

2. <u>Stress Productivity Improvement Incentives</u>

The DOD recognizes the value of plant-wide technology improvement through its TechMod Program. Applications of TechMod funds, however, are both extremely limited and confined generally to prime defense contractors, linked as they are to specific weapons programs. The expansion of the TechMod concept, and the inclusion of the machine tool builders themselves as potential recipients for program funds, would be an efficient way of supporting the viability of the domestic industry through R&D.

Action: DOD should create productivity improvement incentives within the machine tool industry in the form of a TechMod program for machine tool builders that sell to the defense industry and to the DOD itself.

3. Simplify Contracting Procedures

There have been numerous instances where firms, including machine tool firms, have deliberately avoided opportunities to bid on government contracts or apply for government grants because of the complexities of contracting procedures. Such a situation deprives the taxpayer both of greater competition in government contracting and of the savings from reduced bureaucratic requirements.

A reasonable goal in dealings with private firms should be to take no more time in the contracting process for particular items than the average time taken in regular non-government business. With regard to making contract specifications more realistic, and improving disclosure, the procedures used for the recent Watervliet Arsenal FMS procurement might serve as a model.

Action: DOD should consult with the National Machine Tool Builders' Association to establish a program encouraging individual machine tool firms to bid directly for government contracts. Such a program might concentrate on contract specifications (e.g., substituting performance or capacity criteria for design specification criteria), disclosure (e.g., making contract review procedures more open), compliance (e.g., supplying consulting services, through the NMTBA, on EEO, set-asides, etc.), and timing (e.g., stipulating deadlines for reviews and automatic approval if no negative finding is forthcoming by a specific date).

4. Improve Information Flows

If well informed about available R&D funds, machine tool companies with the necessary resources and determination will welcome the chance to improve their technological capabilities. If better informed about manufacturing technologies of interest to DOD, contractors as well as suppliers can respond with more aggressive efforts at plant modernization.

These information flows are especially important given the U.S. machine tool industry's present relatively fragmented structure. As a rule, only the largest machine tool firms have been able to maintain the close relations with university engineering departments, and separate R&D divisions, which are needed to maintain a technological edge.

Action: Establish, in conjunction with the U.S. machine tool industry, one or more joint, industry-wide research centers.

Equip DOD research centers to make a more aggressive effort to make manufacturing technology information available directly to potential adopters. Involve potential adopters in the R&D contract award process.

To increase awareness of ManTech and TechMod activities among process and equipment suppliers, hold regular briefings for suppliers of equipment to acquaint them with the workings of ManTech and TechMod.

Hold process technology forecasting sessions with key individual interest groups (FMS suppliers, near net shape suppliers, etc.) to share with them DOD experts' assessments of related developments taking place in primes that are sponsored by contract R&D money.

5. Require Long-term Production Equipment Maintenance Guarantees

In many cases, one cannot think of "fixing" a machine in the old sense of the word; "repair" has today, in many cases, become the installation of a highly complex circuit board or an electric component made solely by a manufacturer under highly controlled circumstances. Such conditions obviously present enormous challenges even for domestic manufacturers in peacetime, but for overseas resources under wartime conditions, such challenges may be beyond the meeting.

Action: In defense contracts, require that contractors be able to maintain the production equipment for five years even if supply lines are disrupted. Continued production could be guaranteed either by having sufficient parts inventory in the continental United States or by having the ability to replicate the equipment.

6. Study Effects of Consolidation, Acquisitions, and Joint Ventures

The Committee's interest in consolidations, acquisitions, and joint ventures is twofold. First, several countries have adopted policies permitting joint activity that, if engaged in by U.S. companies, would appear contrary to the intent of the U.S. antitrust laws. Such policies place American machine tool manu facturers at a competitive disadvantage. The nation and its lawmakers may have to relax their fears of greater size and concentration of, and coordination among, domestic companies in recognition that many U.S. products, including machine tools, now compete in a world market.

Second, the restructuring of the machine tool industry has, in some instances, involved the purchase of firms by holding companies or conglomerates. Such developments could adversely affect investment in U.S. machine tool production facilities and their ability to respond to Defense Department requirements.

Action: DOD should commission a study of recent consolidations, acquisitions, and joint ventures within the machine tool industry, with the aim of determining whether (1) such actions strengthen or weaken machine tool production in this country, and (2) foreign firms are taking advantage of the relative freedom afforded by their laws to gain a competitive edge. Where concerns are warranted, DOD should present the information to relevant Executive and Legislative branch agencies.

Each of the above recommendations varies considerably in magnitude of effort and resources. The Committee's deliberations, however, were based on the assumption that the opportunities for manufacturing productivity benefits occur each time the DOD procures weapons, equipment, munitions, or spare parts. The DOD's procurement budget for fiscal year 1982 was $64.1 billion. In addition, DOD and the three services administer revolving and management funds, some of which (e.g., DOD stock funds) carry a large procurement quotient. In fiscal year 1984, outlays from these combined funds will exceed $100 billion.

Savings brought about by increases in manufacturing efficiency can have a compound effect, both from the accumulation of productivity gains and the compound savings on interest costs. A one percent productivity gain in the DOD's procurement in fiscal year 1984 alone could, if it became the base for a new level of productivity, save the Department $2 billion in 1990, with cumulative savings over the seven years 1984–1990 of more than $14 billion.

While these gross totals depend upon assumptions about interest rates and increases in procurement spending which may or may not come about, it nevertheless gives a rough estimate of the large savings that can result from productivity gains, and gives some measure of the resources that can justifiably be devoted to this effort.

Beyond this general level of effort, the Committee felt that attempting to develop useful measures of costs and benefits related to each recommendation was a task well beyond their ability and resources because of the huge variety of technologies and applications involved. Meaningful estimates with acceptable levels of confidence would require large volumes of experience data specific to each application. Even then the benefits of applying improvements in manufacturing technology are often difficult to quantify at early stages of the technology's development.

Another important failing of cost-benefit analysis in this context deserves special mention. The Committee believes that where long-term considerations are paramount, reliance on cost-benefit analysis can be self-defeating. For example, the dollar costs and benefits of becoming internationally competitive in machine tool production are difficult to quantify with any degree of certainty; yet such competitiveness is central to many of the concerns of this report. Indeed, the Committee believes that preoccupation with short-term cost-benefit analysis, to the exclusion of important strategic considerations such as the setting of long range goals concerning output and market share, has brought many U.S. firms to the point where they have lost substantial ground to foreign competitors. The question to ask is the cost of *not* staying internationally competitive.

Recommendations for Agencies with Which DOD Has Frequent Contact

The analysis in the body of this report indicates that the U.S. machine tool industry has been harmed as much by domestic economic policies as by the actions of foreign competitors. Changes in the business cycle have had a marked effect on levels of capital investment, R&D, sales, and profitability—possibly more so than in other industry sectors. Machine tool orders are a "lagging" economic indicator; and this means that the industry needs a sustained economic recovery in order to regain a solid equilibrium. The Committee believes that a healthy macro-economy that provides continuous growth over several years could be the most significant single contributor to a healthy domestic machine tool industry.

1. **Raise the Profile in the Administration and the Congress of DOD Programs That Promote Advanced Manufacturing Technologies**

The evident concern in Congress with manufacturing productivity generally does not appear to be matched by efforts to generate appropriations for programs, such as ManTech and TechMod, which would promote manufacturing technology from a departmental level. The Congress needs to focus attention on programs such as these, which hold some promise for solving in a practical way the problems of manufacturing technology lags in U.S. factories.

Action: Congress should appropriate additional funds for ManTech, TechMod, and similar programs as separate line items in the defense appropriations budget.

2. **Build a Program to Promote Machine Tool Exports**

The Committee believes that participation by the U.S. machine tool builders in world markets is essential both for the economic return and to ensure full awareness of foreign technological developments, productivity, and costs. In other words the global machine tool market is a reality in which U.S. firms must participate in order to ensure competitive effectiveness in domestic markets as well as to expand their sales potentials.

In addition, U.S. policy makers must recognize the mobility of technology. Restrictions on U.S. exports for some machine tool technology in an effort to prevent its use by Eastern Bloc countries is apparently not completely effective because of foreign availability.

Action: The Department of Commerce should cooperate with the U.S. machine tool industry to mount a machine tool export promotion program utilizing the resources of the U.S. foreign-based Consular Corps to identify market opportunities and help U.S. manufacturers gain access to those opportunities. This effort would include establishing market controls, providing assistance in proposal preparation, and, where appropriate, facilitating Overseas Private Investment Corporation (OPIC) financing.

In addition, the government should reduce barriers to the export of machine tools to Eastern Bloc countries in cases where those countries have access to the same technology from other sources. U.S. machine tool builders should be able to export the same types of equipment to the Eastern Bloc that other Western countries are exporting to them.

PROBLEM SYNTHESIS AND RECOMMENDATIONS

3. <u>Bring Machine Tool Industry Considerations Into Other Departmental Programs</u>

    Several federal agencies already have task forces looking at the problems of the U.S. machine tool industry. There is no evidence, however, that existing federal programs for technology development have focused on the economics of the machine tool business itself.

    Action: The Administration should inventory the array of federal programs that are aimed at the problems of manufacturing productivity, with the goal of gaining better coordination among programs and simplifying the process of obtaining federal assistance.

Recommendations for Machine Tool Industry

The Committee recommends that the conventional machine tool industry look beyond government trade policy for solutions to its fundamental problems. To be competitive in today's marketplace, now global in nature, machine tool companies will have to modernize their production facilities as well as stay abreast of advanced technologies in their product designs. They should also recognize that American purchasers of machine tools today have begun to consider foreign suppliers very seriously for more reasons than their lower cost. The American machine tool industry should combat the reputation some companies have built for having a reluctance to be responsive to user preference in machine design and systems, a slow delivery record, and insufficient service.

The changing technology will place increasing value on a full product support orientation as the basis for competition. This support would include customer education; needs analysis, applications engineering, and simulation; greater efforts at competing on the basis of quality; and more aggressive service support. The industry should realize that many of the problems that beset it are the same as several other U.S. industries face. As the industry itself has recognized, many solutions must come largely from the machine tool industry itself; some must be implemented on an individual firm basis.

1. <u>More aggressive application of advanced equipment and processes in machine tool production</u>. These steps are needed to improve product reliability, to reduce

costs, and perhaps just as important, to gain first-hand familiarity with modern production methods.
2. <u>A more active search for new technology</u>. This would include taking advantage of access to information available from DOD, and making aggressive efforts to work with prime contractors in areas of new technology that they have identified as important as well as keeping up with offshore technological advances.
3. <u>A greater willingness to invest in long-term competitive strategies rather than responding only to short-term economic considerations</u>. The Japanese firms that are successful in this country have made their mark by responding imaginatively to customer needs with largely standard products. The new competitive realities demand that U.S. firms must do no less.
4. <u>A new acceptance of joint R&D efforts</u>. This would assist in developing a domestic research capability for nurturing advanced production technology in the mid-1980s and beyond.
5. <u>A more extensive information program</u>. The NMTBA should mount a major program to inform machine tool members of the availability of funds and DOD interest in upgrading the machine tool base in the United States.

It is particularly appropriate that the U.S. machine tool builders maximize the value of the current period of cooperation within the industry for more acting upon real operational issues (e.g., labor relations, investment, R&D), which lend themselves to joint efforts. The industry should take this opportunity to set for itself challenging objectives whose attainment will achieve the worldwide competitiveness that is necessary. The challenge facing the industry is to persist with such an agenda until its objectives are realized.

CONCLUSION

The U.S. machine tool industry displays the characteristics of a mature industry facing pressures to undergo fundamental change. The proper response of government to such change is twofold:

1. The government should continue to aid technological progress and the positive restructuring in the industry. This may mean that from time to time the government will have to look into means to overcome the comparative

PROBLEM SYNTHESIS AND RECOMMENDATIONS 97

       advantages of foreign producers where attributable to hidden subsidies such as antitrust concessions and low interest rates give the foreign competitor an advantage.
2. The government should seek to work more directly with machine tool builders to clarify its policy of promoting rationalization and its intention to assist firms that are willing to adapt to the realities of the marketplace.

The realities of the international marketplace, seen from government standpoint, suggest that the U.S. government cannot wash its hands of the industry's concerns. Indeed, several government agencies, including the Department of Defense, the Department of Commerce, the International Trade Commission, and the Export-Import Bank, have planned initiatives aimed at developing more effective policies for the U.S. machine tool industry.

Free market economics, however, assumes that most problems are not amenable to government-imposed solutions. Sometimes resolution of the problem depends upon changing the attitudes and practices within industries suddenly faced with rapid change. The surveys conducted for this report turned up such a pattern among both machine tool builders and users. The challenge facing policy-makers today is to identify those measures which demand government action, and those which are best left to the industry.

In terms of this report, the most relevant reason for action is simply one of our own national defense. But such an effort will also help improve our whole national productivity and cannot be neglected either. The Committee believes that that argument will come to be of far greater importance to this country than any defense argument.

APPENDICES

APPENDIX A
HIGHLIGHTS OF PHASE I STUDY

Defense Needs and the Machine Tool Industry

In a national security emergency, the availability of production capacity to meet "surge" or "mobilization" requirements is critical; machine tools are an important component of that capacity. Several recent reviews have examined the Defense Department's machine tool reserve and found much of it to be obsolete. Similarly, they have considered the domestic machine tool industry's ability to expand capacity and output rapidly and judged it to be inadequate.

In view of the long lead times characteristic of machine tool design, production, and delivery, a large increase in output would require a substantial investment and take several years to achieve. At a time of financial constraints on present weapons systems procurement programs, investment in creating and maintaining extra machine tool capacity to meet emergencies is highly unlikely. Therefore, it is particularly important that the Department of Defense carry out mobilization planning in consultation with machine tool manufacturers and users. Such planning should concentrate on maintaining existing machines in operation by ensuring the supply of spare parts, identifying critical equipment and its sources, and providing for the conversion of civilian machine tool production capacity to military applications. The issue of self-sufficiency versus reliance on foreign sources should also be confronted.

Because of its important bearing on productivity, production rates, and cost containment, modernization of the DOD and contractor-owned machine tool inventory is a critical element of the defense industrial base revitalization strategy called for by the Defense Science Board, the House Armed Services Committee, and others. Such a program would take several years to accomplish. During that period, presumably, the objective would not be to substitute 1970s state-of-the-art machine tools for outdated equipment but progressively to advance and incorporate in defense production new manufacturing technologies. From the point of view of defense needs as well as the competitiveness of the U.S. industry, therefore, two types of DOD policies assume major importance— procurement policies and programs of technology development, innovation, and diffusion.

Previous reports on the defense industrial base have expressed various concerns about DOD procurement practices particularly relevant to the machine tool industry's response to the need for modernization. First, the policy of cost-plus reimbursement is said to discourage contractors' investment in more efficient plant and equipment. Second, Cost Accounting Standard (GAS) 409, requiring depreciation of contractors' tangible assets to be based on their historical or economical useful lives, may prevent full cost recovery in an inflationary period and thus impede replacement of outdated assets with efficient equipment. At the least, CAS 409 imposes a substantial recordkeeping burden on contractors; however, the recent elimination of the Cost Accounting Standards Board leaves no current mechanism for its revision. Third, various restrictions limit the use of multiyear contracting, which is widely believed to offer maximum economies and encourage participation in defense procurement, not least by producers in industries that, like the machine tool industry, are characterized by sharp fluctuations in civilian demand.

DOD manufacturing technology programs have been criticized, not as impediments to innovation, but as inadequate and, in some circumstances, ineffective. The success of the Air Force in developing and promoting the use of numerically controlled (NC) machine tools in the 1950s has not been repeated. Independent research and development (IR&D) funds are rarely available to second-and third-tier contractors. The Manufacturing Technology program has been funded at levels far below those recommended by the Defense Science Board, among others. Generally, manufacturing technology development and innovation must compete for a share of the procurement budget where the acquisition of finished products has far higher priority.

The Manufacturing Technology program sponsors generic technology in hopes that it will be widely transferred. The Technology Modernization program provides funding to address specific problems in particular plants. The panel-drilling robot at General Dynamics in Ft. Worth, where the Technology Modernization investment is expected to have a five-to-one payback, is often pointed to as an example of the program's success. It is a successful example of stimulating the application rather than the development of technology, however, because most of the technology applied by General Dynamics under the program was already available.

The Domestic Machine Tool Industry

The Phase I committee was constituted to identify the issues that must be raised in a more comprehensive study of the industry's potential contribution to the needs of the U.S. Departmemt of Defense, and to plan such a study in outline. In carrying out this charge, however, the committee has made a set of tentative judgments, on the basis of its members' reading and discussion and their experience in management, business analysis, military procurement, and the machine tool industry.

Capital Investment

Inadequate access to capital is commonly raised as the machine tool industry's fundamental problem. The extreme cyclicality of the domestic market is surely a factor in the tendency of investors to view U.S. machine tool companies as risky places to hazard capital. Some sources cite the additional problem of overconservative managements reluctant to make needed investments in either plant or product development. It is also likely that the many small businesses in the machine tool industry have been hurt by high interest rates over the past few years.

This committee finds much of this description plausible. A domestic financial environment more favorable to capital investment would presumably raise sales of machine tools and other forming equipment. But should the domestic industry be unable to compete in technology, marketing, and service, such an environment might only increase the market for foreign manufacturers. Effective management, with the capacity to grasp new technical and market opportunities, is also important.

Labor

With its highly cyclical market, the machine tool industry in the United States understandably finds it difficult to attract and retain skilled craftsmen in numbers necessary to meet business peaks. As a result, delivery on orders during such periods is slowed, intensifying the effects of the industry's common practice of carrying heavy order backlogs. When demand is high, therefore, many buyers turn to foreign machine tools, which can generally be delivered much more quickly.

Capital investment is one solution to this potential shortage. The adoption of new, more efficient manufacturing technology may well diminish the requirement for machinists, tool-and-die makers, and members of other highly skilled occupations.

Higher wages would presumably go far toward attracting the necessary personnel. One government study in any case disputes the long-term impact of labor shortages, citing such indicators as average weekly overtime hours, quit rates, and relative wages.

Of more long-term significance is the industry's ability to attract the talented engineers, designers, and managers who will develop and manufacture the next generations of tools. Experts in cutting and forming technology, electronics, computerized control systems and their software, manufacturing systems design, and marketing, among other fields, will be needed. Some of these specialists are currently in very heavy demand in "growth" industries, and it may not be so easy to attract them to an industry commonly perceived as heavily cyclical and technologically backward. Again, competitive salaries will have some effect, as will the challenge of working in an industry with technological and management challenges before it.

Management

Some recent studies propose that the machine tool industry's slowness to innovate and lack of aggressiveness in marketing may be due largely to the "fragmented" nature of the industry and the specialized, narrow product lines offered by many of the companies. These factors, it is suggested, militate against adequate investment in innovation and in some ways favor unsophisticated management. The Machine Tool Task Force, for example, says, "Small businesses are typically owned and operated by people who were originally craftsmen and they do not usually employ engineers or other university-trained people. As a result, they are, with some outstanding exceptions, nonparticipating members of the technology-exchanging community." Technological change in machine tools and forming technology, the report says, has been prompted over the past 40 years more by user demands (and government-subsidized development) and technical advances in the supplier industries (notably cutting tool manufacturers) than by independent initiatives in the machine tool industry.

As an explanation of the industry's performance, such an analysis is inviting. In a field whose technological and market horizons are expanding as rapidly as those of the forming industry, it is to be expected that small companies with narrow product lines and experience in producing standard products over long periods of time should miss important opportunities for innovation. However, it should not be forgotten that the industry's sales leaders are fully large enough to afford the technical and management resources necessary to take advantage of new technology and new markets.

Capacity

The existence of large order backlogs and long lead times suggests that capacity is insufficient for peak peacetime needs. If the need for mobilization arises, the industry in its present condition will not have time to respond. Capacity concerns involve types of machines as well as quantity.

During mobilization, the easiest capacity to change to meet defense needs is capacity used for exports. Therefore, a machine tool industry that is competitive in world markets during peacetime should be able to meet mobilization demands. It should also be noted that foreign-owned machine tool plants in this country may be used during wartime to meet U.S. defense needs.

Technology and International Competition

The U.S. machine tool industry's reputation for slowness in applying new technology, and for unreliability in the higher technology product lines, is no doubt a significant factor in its market performance against foreign competitors. The extent to which this reputation is deserved is unclear, but there is evidence that it influences buyers.

The domestic market has a relatively older stock of machine tools and therefore appears rather slow to adopt new process technology, compared to those of other industrial nations. The U.S. machine tool industry's failure to market its products strongly overseas has thus, probably, cut it off from sources of more sophisticated demand than those available at home. If so, it has correspondingly reduced its incentives to innovate.

Nor has the U.S. industry benefited from national research and development organizations, such as those established from the machine tool industries of some other countries (notably Japan, West Germany, and France). Many believe that, especially in Japan, government guidance has been critical to the international success of foreign machine tool industries. In addition, the close working relationships between foreign industry and universities are absent in the United States.

Role of Prime Contractors

Many defense contractors are highly capable of developing their own sophisticated tools. Although individual contractors have often developed sophisticated machines in-house, it has usually been machine tool companies that have built such machines, transforming prototypes into heavy-duty equipment suitable for high-volume production and making more standard models available for purchase. It is this role of technology transfer among defense contractors that may be the most important contribution of the domestic machine tool industry—and the one that would be most sorely missed if the domestic industry were to deteriorate further. It would be undesirable, too, to pass on this role to foreign suppliers, however competitive they might be.

Phase I Committee Recommendations for Further (Phase II) Study

The most prominent aspects of the machine tool industry, so far as this committee's charge is concerned, are (a) the rapid expansion of its technological and market horizons over the past decade or so, and (b) its deteriorating position in the world market, as measured by market share at home and overseas. In outlining a plan for a more comprehensive study of the industry's potential contributions to defense needs, the committee has concentrated on these characteristics.

Such a comprehensive study must begin by setting boundaries on the field of investigation somewhat wider than the machine tool industry's traditional limitation to metal-removing equipment, taking into account new materials and the information technologies of control and systems integration. Then, with such a definition in hand, a further study can assess the health of the industry, and its ability to serve Defense Department needs. The following outline embodies this committee's recommendations as to how such a study should proceed.

I. Industry Analysis

As a first step, the industry and its markets should be identified and characterized.

A. Define the machine tool industry. For purposes of this study, the definition should be broad enough to include not only firms traditionally considered part of the machine tool industry, but also manufacturers of manufacturing systems components (machine holding device, cutting tool, gauging and measuring device, controls, and material handling equipment). Include information integration and such competing industrial shaping technologies as near-net-shape forming. Examine the current structure of the machine tool industry, the changes it is undergoing, and its expected evolution over the next 20 years.

B. Assess the technological and economic trends to which the industry should respond. Most important among these trends is the integration of fabrication, assembly, material handling and storage, production control, and management information systems. New methods of metal-forming and metal-cutting as alternative shaping techniques, and importance of new technical disciplines such as computer control, the merger of electronic controls and mechanical processes, changing cost factors in production, market trends, joint international ventures and exchanges of information, and financial considerations should all be assessed.

C. Group the firms in the machine tool industry according to categories that will aid an analysis of the industry's responsiveness to military needs. Which sectors are most important to the Department of Defense? In which firms is research and development being done? Possible categories include high-volume suppliers, suppliers of high-technology equipment, suppliers of equipment particularly critical to military needs, and custom integrators of manufacturing systems. Consider also which classes of tools are important to the Department of Defense.

D. Assess the reasons why some machine tool companies prefer not to seek Defense Department contracts. E. For industry sectors identified as important to the Department of Defense, conduct case studies of their monitoring of the defense environment and their decision-making processes, to test how each type of company is likely to respond to different DOD initiatives or policies.

II. International Competitiveness

The past performance of the U.S. machine tool industry suggests that the industry is losing some of its ability to compete. A more comprehensive study should investigate the facts of the case and assess and weigh the various contributing factors that have been proposed.

A. Export decline analysis

1. To what extent has recent booming domestic demand favored imports? How have domestic manufacturers responded?
2. Is national export-import policy a significant factor?
3. Do intrinsic cost advantages play important roles in foreign manufacturers' success? If so, what are these advantages and how important are they?
4. To what extent do labor and management practices contribute to the success of foreign manufacturers?

APPENDIX A   109

     5. Are claims of superior quality, higher reliability, faster service, and lower prices for foreign goods based on fact?
     6. Which tools are the primary imports, and which the primary exports?

B. Comparison with key competitors (e.g., Japan) from users' perspective: price, quality, delivery, and reliability.

III. Problem Synthesis

On the basis of items I and II, identify the newly defined industry's fundamental problems (if any), describe potential DOD strategies for assisting in correcting these problems, and identify obstacles to putting those strategies in effect. The following issues may provide lines for this analysis:

A. The influences of government policies in the fields of taxation, antitrust restrictions, manpower training and education, research and development, and restrictions of sales to the "Eastern Bloc."

B. Direct funding of research and development relevant to machine tool technology, in both the machine tool industry and universities, by the Department of Defense.

C. Alternative Department of Defense procurement strategies

     1. Is it possible, and under what circumstances would it be desirable, for the Defense Department to modernize the government-owned portion of the defense industrial base on a continuing and sustained basis?
     2. Can and should procurement regulations be changed to foster the installation of capital equipment of defense contractors?
     3. Should research and development funding be augmented? If so, how should funds be allocated between product and process development? How should they be allocated between universities and industry?
     4. Would formation of a joint Defense Department-machine tool industry committee be an effective group to develop plans for surge and mobilization?

IV. Recommendations

The recommendations will follow from the analysis in part III of this Phase II study, as described above. Likely categories for recommendations include the following:
  A. Business Strategies
  B. Procurement Strategies
  C. Technological Strategies

  1. Product research and development
  2. Process research and development

APPENDIX B

POLICIES OF FOREIGN GOVERNMENTS

JAPAN

Government policy has played an important role in stimulating Japan's machine tool industry. Japan's Ministry of International Trade and Industry (MITI) has described three stages of an industry's development: growth, maturity, and decline. MITI's greatest influence is during the first and third stages—supporting growing industries and cushioning the effects of decline. The Japanese machine tool industry is currently considered (by MITI) to be in the maturity stage. MITI played a major role in helping the industry to reach maturity; however, its influence has diminished considerably in recent years.[1] Thus, although much has been said about the large number and variety of Japanese policies that support its machine tool industry, many of these policies are no longer in effect.

Industrial Planning

As part of its statutory function of identifying and promoting industrial growth, MITI has been authorized to:

- provide funds for modernization
- approve rationalization cartels
- stimulate mergers, joint ventures, and further modernization of equipment
- move domestic firms toward increased specialization and international competitiveness[2]

As an example of moves toward specialization, MITI now requires Japanese firms to discontinue manufacturing types of machines that are less than 20 percent of a firm's total production and where the firm's share of national production is less than 5 percent (excluding machining centers).[3] Thus, Japanese machine tool builders benefit from economies of scale and reduced competition.

Availability of Capital

Japanese firms needing capital for expansion or modernization can draw on a range of incentives and traditional practices going beyond what is available in the United States. These include:

- policies that keep interest rates artificially low for loans to favored manufacturing industries
- a relatively concentrated commercial banking sector, which enables the Ministry of Finance and the Bank of Japan to "ration" credit[4]
- a high rate of domestic saving, helped in part by tax preferences on interest income
- a tradition of close cooperation among government agencies that sets economic priorities and commercial lending agencies
- generous depreciation allowances, including a special accelerated depreciation rate for numerical control (NC) machine tools[5]

Although many of these policies were conceived at a time when capital was scarce and when extraordinary efforts were needed to revive a war-damaged industrial base, the same policies now provide Japan with substantially greater investment incentives than exist in any other OECD country.

R&D Incentives

A quasi-governmental corporation, Flexible Manufacturing System Complex (FMC), involves machine tool builders and others in a large-scale, government-sponsored effort to further the state of the art in manufacturing processes. Although perhaps the most visible, this is but one of a number of government-sponsored research projects involving government laboratories, universities, and industry.

Japanese tax laws allow generous credits for research and development (25 percent of incremental R&D).[6]

Because R&D is typically no more than 5 percent of sales, subsidized R&D cannot account for much of the price differential between Japan and the United States. Subsidized R&D does, however, have one major advantage that is not just financial. The government's involvement in R&D lends strategic directions and legitimacy to R&D work and has helped the Japanese develop NC and electronic discharge machines quickly.

The bicycle and motorcycle race wagering tax also provides direct subsidies to the machine tool industry. Though their extent is unknown, total collections in 1981 from this tax, which is earmarked "for promotion of industries related to bicycles and other machines,"[7] were almost $100 million.[8]

FRANCE

Industrial Planning

France has a long tradition of government involvement in the economy having used a variety of market and non-market tools to promote the national economy while reducing dependence on foreign manufacturers. With the election of the Mitterrand government, France has begun to emulate the Japanese, placing greater reliance on market signals but utilizing various government policies to stimulate targeted growth sectors and sponsoring national research projects.[9] On June 29, 1982, the government announced the creation of a "super ministry" of research and industry, modeled after MITI.[10] This ministry will implement the various tools of French industrial policy, including industry restructuring, subsidies, joint ventures, foreign acquisitions by French firms, and research spending to promote growth industries.

The Ministry of Research and Industry has begun implementing a major restructuring plan for the machine tool industry, expected to last through 1986. The establishment of the French Heavy Machinery Company (MFL), a holding company, was announced in July 1982 and was formalized in September as the first step in this restructuring. MFL currently has two subsidiaries, one devoted to milling machines and one to lathes, each formed by the merger of two machine tool companies. The Ministry of Research and Industry has a development contract with MFL that calls for a 200 million franc investment between 1983 and 1985 to increase its share from 4 percent of world production to 6 percent.[11]

MFL is the first of 3 poles around which the machine tool industry will be regrouped. Of the nearly 150 machine tool companies, the most important will be regrouped around 15 industry leaders through mergers and acquisitions. Such arrangements are expected to increase the international competitive position of French machine tools by reducing R&D and manufacturing costs.[12]

Shortly after its creation, the Ministry of Research and Industry announced the "production plan" to bring together various industries—including machine tool—in a national automation effort.[13] This plan aims for 25 percent growth in process control each year for 3 years.

In addition, the government has called for a drastic cut in imports of NC machine tools, from a 60 percent market share to 30 percent by 1984. This reduction is to be achieved through government contracts and subsidies, leading to an increase in NC machine tool production from 27 percent to 60 percent of total machine tool production by 1985, with total machine tool output doubling by 1985. The government expects firms to commit 5 percent of sales to R&D; in return, the government will award contracts of 200 million francs over the next three years.

R&D Incentives

The French government's 1982 budget plans call for a 37 percent increase in the research and development program from the previous year and a quintupling over the next few years. During this time, in contrast, the total budget will rise only slightly. This, combined with recent nationalization of several high technology companies, means that the government controls approximately 75 percent of R&D.

Several programs exist to assist private firms in R&D. One program, Lettres d'Agrement, is a means to encourage firms to develop and manufacture a product in the national interest. The government provides loan guarantees or low-interest loans to assist the firm, with preference given to priority sectors. Aide au Development is a program to assist firms in commercialization of public and private R&D results. The government provides subsidies for prototypes and pilot plants, as well as loans for 50 percent of the project cost. The loans are repaid only if the project is successful. Other programs promote cooperative R&D between private, government, and university labs.

Three centers for goal-oriented machine tool research were designated by the government in 1982; the Machine Tool Study and the Research Center (GERM), the Agency for Development of Automated Production (ADEPA), and the Mechanical Industries Technical Center (CETIM) will be the sites of the programs. Machine tool companies will be encouraged to take advantage of the technological advances developed at these centers.

FEDERAL REPUBLIC OF GERMANY (FRG)

Industrial Planning

During the past decade, the government of the Federal Republic of Germany (FRG) has become increasingly involved in directing industrial development and change. The lead agency in industrial planning is the Ministry of Research and Technology (BMFT), which was created in 1972. Like MITI, BMFT has encouraged rationalization of industries in structural decline and promoted knowledge-intensive sectors. Like Japan in the late 1950s and early 1960s, the FRG has encouraged mergers, consolidations, and offered grants, low-cost loans, and tax concessions during the late 1960s and early 1970s. However these programs have primarily been designed to benefit declining industries; the Germans have no official priority list of growth sectors to be supported. Priorities are set through market mechanisms, while business response to market changes are guided by an informal system of "concertation" based on input from government, banks, and labor.

The Economics Ministry and the BMFT provide grants to industry for research and development. A variety of research institutes, both independent and university associated, receive government funding. For example, the Technical Research Institute at Aachen is considered by many to be the best machine tool laboratory in the world; 69 percent of its funds for research come from the government—either federal or Lander (states).[14]

The 17–20 Fraunhofer Institutes in the FRG are an important source of industrial research. Fraunhofer Institutes specialize in industrial technology, especially in high growth, advanced technology industries, under contract to companies and government agencies. The government matches the institutes' contract funding with an equal amount to be used for basic research.[15]

Since the early 1970s, a major project to develop flexible manufacturing systems (FMS) has been sponsored by the federal government, involving many institutions and firms and receiving heavy subsidies. Elements of the project include CAD/CAM technology; technology for parts fabrication and assembly methods; and highly flexible manufacturing systems based on machining centers grouped around programmable industrial robots. The federal government plans, coordinates, and funds various R&D projects in academic and industrial labs related to FMS. For example, university research centers at Aachen, Berlin, and Stuttgart have been encouraged to develop a research center for CAM.

The federal government has also implemented the Follow-on Production Technology R&D Program to run through 1984 and probably longer. This program is designed to consolidate the development of FMS research by encouraging utilization of R&D results to batch manufacturing processes. Interest free loans and rapid depreciation are provided by the government to promote installation of R&D results, including applications of industrial robots and automatic controls. Over the long term, the Follow-On Program is intended to encourage the use of various FMS in all plants to create computer-integrated automated factories.

NOTES TO APPENDIX B

1. Houdaille Industries, Inc., "Petition to the President of the United States Through the Office of the United States Trade Representative for the Exercise of Presidential Discretion, Authorized by Section 103 of the Revenue Act of 1971, 26 U.S.C. Section 48 (a) (7) (D)," May 3, 1982, p. 64.

2. Ministry of International Trade and Industry, "The Vision of MITI Policies in 1980s, Summary," Provisional Translation, March 17, 1980, p. 15.

3. National Machine Tool Builders' Association, Japanese Study Mission, "Meeting the Japanese Challenge," September 14, 1981, p. 16.

APPENDIX B

4. Houdaille, pp. 61, 74–75.

5. Comptroller General of the United States, <u>Industrial Policy; Japan's Flexible Approach</u>, U.S. Government Printing Office, June 23, 1982, p. 30.

6. James A.Gray, "America Needs You." NMTBA, p. 2.

7. Comptroller General of the United States, p. 36.

8. Houdaille, pp. 110–111.

9. NMTBA, Japanese Study Mission, p. 16.

10. Several national research projects are in areas related to machine tools, including electronics, computers, and robotics.

11. G.Bidal, "With the 'Productics' Plan, National-Scale Automation is at Stake," <u>Electroniquetualites</u>, Paris, September 3, 1982, p. 1, translated in <u>West Europe Report on Science and Technology</u>.

12. George LeGall, in <u>L'Usine Nouvelle</u>, Paris, October 7, 1982, p. 82, and Lubka, Stephane, "Machine Tools: Official Birth of the French Heavy Machinery Company," <u>Les Echos</u>, Paris, September 3, 1982, p. 6. Both translated in <u>West Europe Report on Science and Technology</u>.

13. <u>AFP Sciences</u>, Paris, July 1, 1982, p. 1, translated in <u>West Europe Report on Science and Technology</u>.

14. Andre Larane, "Aachen, a Mecca for Machine Tools," <u>Industries & Techniques</u>, Paris, June 1, 1982, p. 77.

15. Department of Commerce, Office of Productivity, Technology, and Innovation, <u>Cooperative R&D in Major OECD Countries</u>, June 30, 1982, p. 7.

APPENDIX C

QUESTIONNAIRE SENT TO MACHINE TOOL INDUSTRY EXECUTIVES

As part of a study of the U.S. machine tool industry and defense preparedness, the National Research Council is examining technological and economic trends that affect the industry. We would like your reactions to our list of trends, which follows. Please use the scales to indicate how much the factors affect your firm and the industry as a whole (i.e., now much change you expect these trends to require). We would appreciate any comments on the impact of these trends, how you plan to or currently respond to them, and expectations for the future. There is room at the end of each section to suggest other factors that we may have missed. If you have any questions, call Miss Janice Greene at (202)/334–2570. Please respond by March 18. Thank you for your assistance.

Technological Trends

		Importance to your firm		Importance to the industry
1.	Material substitution (composites, plastics, ceramics)			
	a. in design of product to be machined	( ) ( ) ( )	high medium low	( ) ( ) ( )
	b. in design of machine tools	( ) ( ) ( )	high medium low	( ) ( ) ( )

Comments:_____
_____

2.	New forming techniques to reduce metal-cutting and labor-intensive finishing (near net shape, powder metallurgy, precision casting)	( ) ( ) ( )	high medium low	( ) ( ) ( )

Comments:_____
_____

APPENDIX C

		Importance to your firm	Importance to the industry
3.	New cutting techniques (lasers, electro-deposition, chemical milling, advanced grinding technology)	( ) high ( ) medium ( ) low	( ) ( ) ( )

Comments:_____

4.	Increasing use of automation		
	a. Small batch fabrication	( ) high ( ) medium ( ) low	( ) ( ) ( )
	b. Assembly	( ) high ( ) medium ( ) low	( ) ( ) ( )
	c. Group Technology	( ) high ( ) medium ( ) low	( ) ( ) ( )
	d. Adaptive controls & systems	( ) high ( ) medium ( ) low	( ) ( ) ( )
	e. Computer-integrated design and manufacturing	( ) high ( ) medium ( ) low	( ) ( ) ( )

Comments:_____

5.	Increased procurement of systems from full-line and systems houses	( ) high ( ) medium ( ) low	( ) ( ) ( )

Comments:_____

APPENDIX C

		Importance to your firm		Importance to the industry
6.	Increasing demand for precision machine tool performance and repeatability	( ) ( ) ( )	high medium low	( ) ( ) ( )

Comments:_____
_____

7.	Increasing demand for high pressure high temperature, and high speed capabilities in equipment	( ) ( ) ( )	high medium low	( ) ( ) ( )

Comments:_____
_____

8. Other:_____
_____

Economic Trends

		Importance to your firm		Importance to the industry
1.	Increasing competition from foreign manufacturers in both domestic and foreign markets	( ) ( ) ( )	high medium low	( ) ( ) ( )

Comments:_____
_____

APPENDIX C

		Importance to your firm		Importance to the industry
2.	Increasing concentration of the industry through mergers, takeovers, and failures	( ) ( ) ( )	high medium low	( ) ( ) ( )

Comments:_____

_____

3.	Increasing relations with foreign firms, through licensing, joint ventures, and acquisitions	( ) ( ) ( )	high medium low	( ) ( ) ( )

Comments:_____

_____

4.	Diversification and integration with related industries	( ) ( ) ( )	high medium low	( ) ( ) ( )

Comments:_____

_____

5.	Current low liquidity and flexibility of the industry, leading to reduced R&D and a shortage of working capital	( ) ( ) ( )	high medium low	( ) ( ) ( )

Comments:_____

_____

APPENDIX C

		Importance to your firm		Importance to the industry
6.	High cost of capital over a prolonged period	( ) ( ) ( )	high medium low	( ) ( ) ( )

Comments:_____

_____

| 7. | Unprecedented unpredictability of economic conditions (regarding growth, inflation, interest rates, currency exchange rates) | ( )
( )
( ) | high
medium
low | ( )
( )
( ) |

Comments:_____

_____

| 8. | Factory utilization (lengthy decline, followed by recent upturn) | ( )
( )
( ) | high
medium
low | ( )
( )
( ) |

Comments:_____

_____

9. Other:_____

_____

Name and _____
Address: _____
_____